ESTABLISHING MEDICAL REALITY

Philosophy and Medicine

VOLUME 90

Founding Co-Editor
Stuart F. Spicker

Editor

H. Tristram Engelhardt, Jr., *Department of Philosophy, Rice University, and Baylor College of Medicine, Houston, Texas*

Associate Editor

Kevin Wm. Wildes, S.J., *Department of Philosophy and Kennedy Institute of Ethics, Georgetown University, Washington, D.C.*

Editorial Board

George J. Agich, *Department of Bioethics, The Cleveland Clinic Foundation, Cleveland, Ohio*

Nicholas Capaldi, *Department of Philosophy, University of Tulsa, Tulsa, Oklahoma*

Edmund Erde, *University of Medicine and Dentistry of New Jersey, Stratford, New Jersey*

Eric T. Juengst, *Center for Biomedical Ethics, Case Western Reserve University, Cleveland, Ohio*

Christopher Tollefsen, *Department of Philosophy, University of South Carolina, Columbia, South Carolina*

Becky White, *Department of Philosophy, California State University, Chico, California*

ESTABLISHING MEDICAL REALITY

ESSAYS IN THE METAPHYSICS AND EPISTEMOLOGY OF BIOMEDICAL SCIENCE

edited by

HAROLD KINCAID

University of Alabama at Birmingham, Birmingham, Alabama, USA

and

JENNIFER McKITRICK

University of Nebraska-Lincoln, Lincoln, Nebraska, USA

 Springer

A C.I.P. Catalogue record for this book is available from the Library of Congress.

ISBN-10 1-4020-5215-4 (HB)
ISBN-13 978-1-4020-5215-6 (HB)
ISBN-10 1-4020-5216-2 (e-book)
ISBN-13 978-1-4020-5216-3 (e-book)

Published by Springer,
P.O. Box 17, 3300 AA Dordrecht, The Netherlands.

www.springer.com

Printed on acid-free paper

TABLE OF CONTENTS

ACKNOWLEDGEMENTS

We would like to thank the Center for Ethics and Values in the Sciences at the University of Alabama at Birmingham for financial support and Pam Williams, Minnie Randle, and Lisa Rasmussen for editorial and logistical support. Chapter 8 is reprinted with permission from Robert Perlman, Why Disease Persists: An Evolutionary Nosology, *Medicine, Health Care and Philosophy Volume 8, Number 3/November, 2005.*

1. INTRODUCTION

Medicine has been a very fruitful source of significant issues for philosophy over the last 30 years. The vast majority of the issues discussed have been normative – they have been problems in morality and political philosophy that now make up the field called bioethics. However, biomedical science presents many other philosophical questions that have gotten relatively little attention, particularly topics in metaphysics, epistemology and philosophy of science. This volume focuses on problems in these areas as they surface in biomedical science.

Important changes in philosophy make biomedical science an especially interesting area of inquiry. Contemporary philosophy is largely naturalistic in approach – it takes philosophy to be constrained by the results of the natural sciences and able to contribute to the natural sciences as well. Exactly what those constraints and contributions should be is a matter of controversy. What is not controversial is that important questions in philosophy of science and metaphysics are raised by the practice of science. Physics, biology, and economics have all drawn extensive philosophical analysis, so much so that philosophical study of these areas have become specialized subdisciplines within philosophy of science. Philosophy of medicine approached from the perspective of philosophy of science – with important exceptions (Schaffner, 1993; Thagard, 2000) – has been relatively undeveloped.

Nonetheless, medicine should have a central place in epistemological and metaphysical debates over science. It is unarguably the most practically important of the sciences. It also draws by the far the greatest resources and research efforts of any area in biology. Yet philosophy of biology has focused almost exclusively on evolutionary biology, leaving the vast enterprises of immunology, cancer biology, virology, clinical medicine, and so on unexplored. Naturalized philosophy has emphasized the important interplay of historical and sociological aspects of science with its philosophical interpretation. Biomedical science as a large scale social enterprise is a natural target for such approaches. Relatedly, within philosophy there has been a growing interest and appreciation for the connections between issues of value and issues of fact in science (Kincaid et al., 2007). Biomedical science is a paradigm instance where the two intersect.

The upshot is that biomedical science is a potential rich area for philosophical investigation in areas outside biomedical ethics. This volume seeks to show that promise and to encourage its exploration.

Aside from this general naturalist philosophical perspective which all the papers in this volume share, a number of more specific themes emerge. Not surprisingly, given the central place that the concept of disease plays in medicine, the status of natural kinds is a recurring concern. Yet that concern surfaces in both a practical

1

H. Kincaid and J. McKitrick (eds.), Establishing Medical Reality, 1–11.
© 2007 *Springer.*

and skeptical way: practical in that concrete issues such as the nature of genetic disease or the reality of mental illness are at the fore, skeptical in that the notion of a natural kind itself is up for debate. A second recurring theme is the contestability of medical knowledge. None of the contributors argues that medical knowledge is a mere construct, but many are sensitive to the complex relations between data, conclusions, and practical necessities that exist in medicine. A related and third important theme cutting across the essays is that philosophy can learn from what other disciplines have to say about medicine. Medicine has important connections to other biological and social sciences; philosophy can gain much from interacting with other disciplines such as sociology in trying to understand how medicine works. A corollary here of course is that philosophical understanding of general issues in metaphysics and philosophy of science has much to learn from medicine and other disciplines as well. We expand on these themes below, as a preview to their extensive development in the essays to follow.

The interrelated questions 'what is health?' and 'what is disease?' are central to the philosophy of medicine. The answers to these questions bear on a number of questions related to medicine. People disagree about which conditions count as diseases, and this has implications for what happens in a health care system. The traditional role of medicine is to fight disease and restore health. The purpose of health insurance, both public and private, is to assist those who suffer from disease. So, it seems important to have a clear idea of just what it means for something to be a disease. A key issue in debates over how to define disease has been over the role of values in the analysis. While taking a broadly naturalistic approach, normativists claim that one cannot adequately define disease without employing value-laden concepts, such as harm, suffering, undesirability, and disability. Naturalists, on the other hand, claim to define disease without appeal to values, by employing purportedly descriptive concepts like abnormality and biological dysfunction. Four papers in this collection carve out four different positions in this debate.

In 'Normality, Disease, and Enhancement,' Theodore Benditt joins the naturalist tradition in claiming that disease can be defined non-normatively, in terms of abnormality. He considers several challenges to this view, including the idea that someone can be 'different' without thereby being in need of medical treatment. His response is two fold. He limits the types of abnormality that count as disease, but admits that his view will count more things as diseases than some would like. However, he distinguishes the question of whether something qualifies as a disease from the question of whether someone should do something about it. Even if something qualifies as a disease by Benditt's standards, the practical implications for medicine are unclear. In this sense, he is skeptical about the concept of normality determining the physician's role. Another role that the concept of normality might play in medicine is rejected by Benditt as well. Whether or not one thinks that a physician should always restore normal functioning, one might think that a physician should do *no more than* restore normal functioning – in other words, that doctors should at most cure diseases, not provide enhancements. Benditt compares this issue to the role of 'normality' and 'enhancement' in athletics. It's common to think that athletic competition should be between normal, rather than enhanced, participants. While Benditt finds reasons to limit enhancements in athletic competitions following from the reasons competitions are held in the first place, he finds no comparable reasons to limit medical enhancements generally. Consequently, while the concept of normality plays important roles in some areas of life such as athletics, Benditt claims it is largely misplaced in medicine.

In 'Holistic Theories of Health as Applicable to Non-human Living Beings,' Lennart Nordenfelt defends the normative or holistic approach against naturalist, bio-statistical approaches. According to the bio-statistical approach, a disease is an internal state which is an impairment or environmentally caused limitation of normal functional ability below typical efficacy, decreasing potential for survival and reproduction. Nordenfelt, on the other hand, defines disease as a bodily or mental process which tends to cause an illness – a state of suffering or disability experienced by the subject. While health may be thought of as absence of disease, it is better described on this view as a state in which a person can realize his or her vital goals (states necessary for minimal happiness in the long run) given standard circumstances. Of these two contrasting approaches, it seems as though the bio-statistical approach would have wider applicability than a holistic one. While medicine is primarily concerned with the health of people, the concepts of health and disease extend to non-humans – animals and even plants. The bio-statistical approach aptly accommodates the application of the concept of disease to non-humans. Determining whether a plant or animal is unhealthy would be merely a matter of determining whether some organ or structure was performing its normal function with typical efficacy. It is less clear how to apply the holistic theory of health to non-humans. Nordenfelt's approach defines health and disease in terms of the experiences of suffering or happiness, and many non-human organisms, most notably plants, are by all accounts, not capable of having these experiences. So, defining health and disease for non-humans presents a major challenge to holistic accounts as against bio-statistical accounts. Nordenfelt tries to meet this challenge by weakening the psychological components of the accounts of illness, health, and disease. However, his main claim against the naturalists seems to be that most organisms can have various interests (things that are in their interest, not necessarily things they're consciously interested in) beyond survival and reproduction, and to say that only internal states which threaten survival and reproduction count as illnesses for any kind of organism is too narrow a view.

What if the normativist is right, and the concept of disease is essentially normative? What does that entail about the nature of disease? If diseases are normative, are they out in the world for doctors and biologists to discover, or do people invent or create diseases based on their values? If one is a moral realist, one can maintain that certain things are just bad for certain organisms, regardless of anyone's perception of them as such. This, in turn, would allow the normativist to be a realist about disease. However, if one thinks that values only arise when someone values something, whether an organism is diseased or healthy would also depend on what people value. If cultures differ with respect to disvaluing homosexual desires, deafness, or goiters, for example, whether these conditions are diseases would vary from culture to culture. On this view, diseases are constructed rather than discovered, and do not exist independent of culture.

In 'The Spread of Disease: How to Understand and Resolve a Dispute about the Reality of Disease,' Robert D'Amico defends realism about disease against this social-constructivist approach. While he concedes to the normativist that medical concepts such as 'health,' 'illness,' 'pathology,' 'suffering,' and 'harm' are essentially normative, he argues that the concept of 'disease' is a special exception, and hence not subject to the relativization of other normative concepts. (Hence, D'Amico is a naturalist with respect to diseases, but not with respect to medical concepts generally.) Social constructivists argue that disease states have more in common with aesthetic or moral phenomena than with natural phenomena. Unless one is a thorough-going anti-realist or social constructivist, whether something is water, gold, or an acid is determined by something beyond the mind of any human perceiver. However, according to the social constructivist, whether something is ugly, bad, *or sick*, depends on the social group which makes this determination. D'Amico draws on concepts from the philosophy of mind and the philosophy of science, namely the distinction between strong and weak supervenience, to redescribe this debate. According to D'Amico, the social constructivist's claim ultimately comes down to the idea that while diseases may weakly supervene on physical states of organisms, they do not strongly supervene on such states. After casting his target in these terms, D'Amico marshals arguments to the contrary. For instance, it is reasonable to think that if a person has kidney disease, then necessarily, any perfect duplicate of that person, no matter the social environment, would have kidney disease. That is to say, there can be no difference in disease-state without some difference in physical state – strong supervenience holds.

While D'Amico and Benditt defend naturalist approaches to analyzing disease and Nordenfelt defends normitivism, Peter Schwartz challenges the distinction itself. In 'Decision and Discovery in Defining 'Disease,' Schwartz claims that a more meaningful classification of views is in terms of value-requiring and non-value-requiring on the one hand and dysfunction-requiring and non-dysfunction-requiring on the other. So, for example, a view that requires a disease to be a dysfunction need not be non-normative. But more importantly, Schwartz questions the utility of any of these distinctions. He charges that philosophers who offer analyses of disease are struck in the outdated analytic project of providing necessary and sufficient conditions for the application of a concept. Once you claim to have given necessary and sufficient conditions for having a disease, you open yourself up to the trench warfare of counterexamples, struggling to deal with variety of bodily states which defy your favored classificatory scheme. However, it has long been acknowledged that ordinary concepts are not adequately captured by necessary and sufficient conditions. Ordinary concepts admit of varying degrees of applicability, with better, worse, and borderline examples. Arguably, when people call a condition a disease, they are not checking the condition against a list of criteria which it must satisfy. And quite plausibly, there is no one uniform concept of disease that we all share.

So, rather than give an analysis of the ordinary concept of disease, perhaps what philosophers of medicine should do is to stipulate a particular concept of disease that will be useful to philosophy of medicine or to medicine itself. Schwartz suggests that the dysfunction-requiring accounts, while not an analysis of all or event most concepts of disease, can provide a useful explication of disease.

Assuming that disease exists, and that we have some idea what it is, we can further explicate the notion by breaking it down into genetic diseases, racially-linked diseases, psychological disorders, and many other sub-categories. However, some question the reality of these categories used to understand disease. Do these categories correspond to real distinctions in nature? Classifications of conditions as mental disorders are particularly susceptible to criticism that they merely reflect societal expectations, as the essays of McKitrick and Horwitz suggest (see below). Other essays in this volume explore related ontological/classificatory questions. Is it possible or meaningful to classify some diseases as genetic and others as non-genetic? Or does doing so reflect a misguided either/or attitude towards genetic determinism? While the category of race has been eliminated from most scientific contexts, can we still establish meaningful correlations between race and disease?

In 'Race and Scientific Reduction,' Mark Risjord notes that epidemiological research has identified a number of important correlations between race and health. Blacks suffer higher rates of HIV infection, diabetes, hyper-tension, and cardiovascular disease than whites. Since race has been eliminated from biology, this data is philosophically puzzling. If epidemiology and the other health sciences are based on biology, one would expect race to be eliminated from these disciplines too. Risjord sheds light on this puzzle by reflecting on scientific reduction and elimination more generally. Concepts like heat can be reduced to more fundamental scientific concepts, whereas concepts like phlogiston are better eliminated altogether. Should the concept of race be reduced, eliminated, or neither? According to 'promiscuous realism,' the study of race in medicine and the social sciences is ontologically autonomous – race exists as a social or medical entity, but not as a biological one. Against the promiscuous realists, Risjord argues for a causal-explanatory criterion of ontological commitment. Since, according to Risjord, race does not meet this criterion, race has rightfully undergone an ontological reduction, or elimination, at all levels. Race does not exist. However, questions about race and racial discrimination at the social and medical levels remain important. Risjord retains the role of the concept of race in these realms by distinguishing between epistemic and ontological reductions. While race has been ontologically reduced, it has not been epistemically reduced. This distinction allows Risjord to maintain the seemingly paradoxical conclusion that the study of race in medicine and the social sciences remains autonomous of biology, but neither medicine, sociology, nor biology are committed to the existence of race.

In 'Towards an Adequate Account of Genetic Disease,' Kelly Smith asks 'What exactly is a genetic disease?' For a phrase one hears on a daily basis, there has been surprisingly little analysis of the underlying concept. Medical doctors seem perfectly willing to admit that the causal origins of diseases are typically complex, with a great many factors interacting to bring about a given condition. On such a view, descriptions of diseases like cancer as genetic seem at best highly simplistic, and at worst philosophically indefensible. On the other hand, there is clearly some practical value to be had by classifying diseases according to their predominant cause when this can be accomplished in a theoretically satisfactory manner. The question therefore becomes exactly how one should go about selecting a single causal factor among many to explain the presence of disease. Smith argues that previous attempts to defend such causal selection have been clearly inadequate. In a spirit similar to Schwatz's advocation of explicative stipulation, Smith proposes an epidemiological account of disease causation which walks the fine line between practical applicability and theoretical considerations of causal complexity and attempts to compromise between patient-centered and population-centered concepts of disease. Smith claims that the epidemiological account is the most basic framework consistent with our strongly held intuitions about the causal classification of disease, yet it avoids the difficulties encountered by its competitors.

A further question about genetic and other diseases is 'why do they persist?' If evolution favors those who resist disease, one might expect subsequent populations to be less and less susceptible to disease. However, there is no evidence that this has happened or is happening. In 'Why Disease Persists: An Evolutionary Nosology,' Robert Perlman explains why. Natural selection is not the only process that changes gene frequencies in populations; mutation and other processes may introduce or increase the frequency of genetic diseases. Interactions between genes complicate the relationship between genotype and phenotype, and may result in the preservation of genetic diseases.

Variations in environmental resources and random developmental events further complicate the genotype–phenotype relationship and may also lead to disease. Natural selection increases fitness, but the declining force of natural selection with age is only one indication that fitness is not equivalent to the absence of disease. Natural selection acts on genes, cells, and groups, as well as on organisms; the outcome of evolution reflects selection at different levels of biological organization. Finally, the human environment is continually changing, largely because of the evolution of our parasites and because of changes in cultural beliefs and practices; genetic evolution is comparatively slow and lags behind environmental change. An evolutionary study of disease complements the traditional medical approach and enhances our understanding of the persistence of disease.

Regardless of which approach to defining diseases one takes, there will still be further tough questions about specific cases – arguments over whether some purported disease really is a disease or whether it is properly understood. The papers by Horowitz and McKitrick raise such issues about mental illness.

Horwitz in previous work had used a hybrid notion of disease as a harmful dysfunction to argue that much of what we call mental illness – e.g. much in DSM IV – are not diseases but problems in living. In 'Creating Mental Illness in Nondiosrdered Communities,' Horwitz looks at assessments of the prevalence of mental illness. Studies claim to show that one in four Americans will have a major depressive episode in their lifetime and one of eight, social anxiety disorder. Yet the latter was described as 'rare' in DSM III in 1980. These estimates of prevalence are done by strictly adhered to survey questions analyzed by computer. The surveys are unable to screen out cases where symptoms result not from underlying disease but from transient social stresses – depression due to the loss of a loved one gets categorized with depressive episodes of bipolar disorder; understandable nervousness for someone who rarely speaks in public gets categorized with someone who can never be around more than a couple people. There are vested interests – the NIMH and patient advocacy groups – who benefit if the estimates of mental illness are high and thus the survey process continues. In this sense mental illness is 'socially constructed,' for the labeling process is indeed a sociological one. Horwitz holds that this is compatible with taking some mental illness to be fully objective diseases.

In 'Gender Identity Disorder,' McKitrick examines a condition labeled a mental illness according to DSM IV. GID is generally thought to occur when a patient has a gender identity that is atypical for his or her biological sex. Gender identity is thus not determined by biological sex. Nor is it, McKitrick argues, determined by psychological properties – being effeminate does not mean a man is woman in a man's body. Thus GID assumes there is an essence of gender identity that is independent of biological, psychological, and social characteristics. McKitrick takes this presupposition to be quite implausible, for if two individuals are identical biologically, psychologically, and socially, then they ought to have the same gender (gender should supervene on these traits). The solution to the problem is to see that gender identity has no essence. Gender concepts are cluster concepts that cannot be defined in terms of necessary and sufficient conditions. This does not mean that gender identity is a mere fiction. People do identify with different genders. Yet gender identity is a cluster concept that varies according to differing gender norms, not an essence. McKitrick concludes that the DSM IV definition really makes GID a matter of nonconformity to gender roles, something it denies. Recognizing that GID is a social conflict, not a disorder, calls for greater tolerance of social differences, not for treatment of an illness.

The remaining essays are concerned with the nature and role of evidence in biomedical science. There is now an exhaustive body of work in the philosophy of science showing that the relations between evidence, theory, explanation, and the social processes of science are varied and complex. These essays certainly second that opinion for the case of biomedical science.

In 'Clinical Trials as Nomological Machines,' Bluhm raises interesting questions about the gold standard for evidence in contemporary medicine, namely, the randomized clinical trial. Every drug approved by the US FDA must show efficacy in two Phase III clinical trials where the treatment is compared to outcomes in a control group that gets either a placebo or sometimes a competing treatment. Bluhm's main question concerns what success in these trials tells us about what should be done in clinical practice. Bluhm's conclusion: they tell us much less than is commonly claimed.

Her argument relies on two concepts from the philosophy of science. The first is the idea of nomonological machine. Cartwright argues that experiments in the natural sciences are about realizing a set of specific causal processes that are shielded from outside influences. Only in such cases, according to Cartwright, are we to expect science to find laws in the sense of universal regularities. Outside of the experimental setup we only have knowledge about capacities, not regularities. Clinical trials, Bluhm argues, are attempts to establish nomonological machines – to provide a situation where interfering factors are sufficiently controlled that we can have reliable knowledge that the treatment caused the observed differences in outcomes.

The problem for the clinician is trying to tell what the experimental outcome tells us about how the treatment will work for the individual patient. There is a widespread tendency to think a successful RCT shows the real effect of the drug. However, it in fact only tells us about the drug's capacity – what it can potentially do – in situations *outside the experimental set up* and it does not give us very much information about that capacity. Clinicians thus must 'reverse engineer' the experiment. 'Reverse engineering' is Dennett description of how evolutionary biologists must go about trying to identify function in biological organization – by asking how the observed outcome might have been put together. For clinicians to apply RCT results to their own patients, they need to know all the decisions that went into designing the clinical trial in order to make a reasonable judgment about whether the results are relevant to their patient. Published results, however, seldom provide the needed information.

In 'The Social Epistemology of NIH Consensus Conferences,' Miriam Solomon also applies ideas from recent philosophy of science to a fundamental epistemic practice in medicine. The practice is the NIH Consensus Conference. These are designed to influence the behavior of physicians by bringing together unbiased experts to reach objective recommendations based on what the evidence shows in a particular area of medicine where there currently is controversy and a gap between theory and practice. The ideas employed from recent philosophy of science concern the many different ways that bias can surface in science as revealed by many historical and case studies. Like Bluhm, Solomon reaches the conclusion that a gold standard in medicine – this time, the Consensus Conference – is not nearly as effective as advertised.

There are two fundamental questions about consensus conferences: do they produce objective decisions on controversial treatment decisions and do they change the medical practice of physicians on the ground? Solomon argues that they fail in the first goal for several reasons. Consensus conferences are preceded by government conducted metaanalysis of relevant studies and these results are announced before hand. Thus in practice consensus is often reached before the conference, not by the conference itself. Moreover, there is good evidence that disciplinary biases – biases that everybody in a given area share – are not weeded out. For example, a recent conference on dental treatments reached entirely different conclusions than the proceeding metaanalysis, but gave no explanation why; the most likely explanation is that the metaanalysis failed to show that standard dental practices are effective, threatening the interests of dentistry. Finally, obvious checks on the reliability of consensus conferences have not been done – no one has checked to see how frequently their recommendations hold up to further scrutiny.

Solomon concludes that consensus conferences primarily serve a rhetorical function – persuading physicians who are unlikely to be persuaded by metaanalyses. Solomon provides a variety of evidence showing that they are not very effective in achieving this goal as well and that when they do so, it is by changing reimbursement schemes rather than providing a seemingly objective analysis of the evidence.

Moira Howes in 'Maternal Agency and the Immunological Paradox of Pregnancy' continues the theme of applying philosophical scrutiny to biomedical evidence. Her target, however, is not a generalized form of evidence as in Solomon and Bluhm. Rather, she focuses on accounts of the immune system in fertility and pregnancy. This has clear practical import, for immune explanations and treatments are gaining in popularity but have undetermined efficacy.

Howes challenges two assumptions in explanations of the immune system's role in fertility: the ontological assumption that mother and fetus are separate entities and the description of their relation primarily in terms of conflict. Science studies has shown again and again that many assumptions go into the interpretation and explanation of data, assumptions that are sometimes underdetermined by the evidence and that reflect social values of scientists. Howes argues for a similar conclusion regarding the standard picture of maternal–fetal relations. That picture originated in large in Medawar's 'immunological paradox of pregnancy.' The paradox arises because according to Medawar the fetus is the immune equivalent of an organ transplant – it is a foreign body. Thus there must be in pregnancy various mechanisms to ward off an immune attack on the fetus; herein lies the relevance of the immune system to infertility. Other widespread analogies for pregnancy are the invasion of cancer cells or of a parasite.

This picture, Howes argues, is permeated with metaphors that reflect larger social values. There is much biological reason to think that the mother and fetus are not two separate entities in the way that a host and a parasite are. For example, they exchange cells that remain in each other for life. Pregnancy is not a disorder in the way the cancer analogy implies nor is the female body merely passive in the process of pregnancy as the invasion metaphor suggests. The parasite metaphor ignores the genetic ties between mother and fetus and the strong evolutionary reasons to think the mother's body has a strong interest in protecting the fetus. Dropping the metaphors of foreign body, parasite, and invasive disease might help reproductive immunology to develop explanations that are truer to the real biological complexity of pregnancy and infertility.

The last two papers by Kaplan and Gifford are the only two in the volume directly concerned with policy and ethical questions. Yet in both cases it is the kind of complexities in biomedical evidence discussed by the other authors that drives the normative questions. Kaplan evaluates public health approaches to violence, Gifford, the ethical legitimacy of clinical trials.

Violent crime is a serious social problem internationally. In all cases a large proportion of the crimes are committed by a few young males. This fact has led many to suggest that there is a biological explanation for crime and that identifying that biological basis would provide useful information about how and when to intervene to reduce crime. In 'Violence and Public Health' Kaplan rejects this conclusion. While he doubts that we currently have any very well confirmed understanding of the biology of crime, the main thrust of his argument is that even if we did, it would not provide useful information for preventing crime. Kaplan identifies two broad biological approaches to violent crime. One seeks to find biochemical differences between those who do and do not commit crimes or, in a more complex mode, biochemical differences interacting with social differences. Deficiencies in MAOA (monoamine oxidase A) in combination with abusive family environments is one such explanation that has gotten serious attention. The second approach uses evolutionary psychology to treat crime as an adaptive response that is triggered in response to environments with low life spans and low levels of social advancement.

While these explanations are intellectually interesting and might some day be well confirmed, Kaplan argues that they are not relevant for what we should do about crime. At present we have no biochemical correlate that predicts crime and no way to alter MAOA levels even if they were well correlated with criminal behavior. We also have only speculative reasons to think crime results from a genetically based strategy that maximizes reproductive fitness. But we do know that various social interventions can be effective, that the environments they target would be key explanatory variables in any biologically based account, and that the results they produce are generally good things for other reasons as well. Given this, focusing primarily on the biological basis of crime is an irrelevant and potential damaging distraction. Thus Kaplan illustrates well a theme mentioned earlier, namely, the complex interplay between value issues and evidence in biomedical science.

Finally, in 'Taking Equipoise Seriously' Gifford is concerned about the interplay between different degrees of evidence and the ethics of conducting clinical trials. The basic problem is this: the treating physician has an obligation to provide the patient with the best possible care. However, in clinical trials there may be some evidence that one treatment is better than another long before the trial is finished. So how can a physician ethically enroll a patient to be randomly assigned to either treatment or control if there is evidence that one arm is better? A standard and widely invoked solution Freedman's (1987) notion of 'clinical equipoise.' So long as the community of practitioners has not reached consensus on which treatment is best, enrolling patients in a trial is justifiable because it is not known which treatment is favored.

Most of Gifford's chapter consists in showing that this widely accepted answer is inherently ambiguous and that as a result the appearance of a solution to the ethical dilemma is misleading. Gifford identifies three different sets of variables that have to be specified to make Friedman's claim unambiguous: (1) community can either be taken in the narrow sense of experts or in the broad sense of physicians at large, (2) consensus can be given either a strict definition as unanimous agreement or a weaker definition as preponderance, and (3) community opinion might be on the decision whether the treatment should be approved for general use vs. the question of whether I would want to use it given what is known now. The latter distinction is motivated by the thought that we have higher standards of evidence for what will be generally used by the public than we do for decisions only affecting a single individual, echoing the role of consensus conferences discussed by Solomon.

With these distinctions in hand, Gifford argues that Friedman's equipoise solution is not supportable. One rationale given for that proposal is that individual physicians ought to take into consideration the views of the community, for that is evidence in itself. However, that justification supports using the preponderance of opinion reading of consensus. But if equipoise is disturbed as soon as there is a preponderance of community opinion, that can happen early in a clinical trial as the result comes in. So standard practice is still not justified. Moreover, even when there is community consensus in either sense about what to give an individual, there need not be consensus about whether the treatment in question should be publically available. Yet it is the latter question clinical trials are trying to answer. Gifford concludes that there is no easy solution and that we need an explicit discussion of how to trade off various values in conducting clinical trials.

We end by noting one last common element to most of the essays, one visible in Gifford's conclusion. While various arguments and theses are found implausible by the authors, most do not claim to have the final word on how the various problems discussed should be decided. Simplistic solutions can be rejected, but more careful discussion and investigation is called for to handle the problems raised. That is a strong indication that biomedical science is a fertile ground for ongoing work in applied epistemology and metaphysics.

University of Alabama at Birmingham, Birmingham, Alabama, USA
University of Nebraska, Lincoln, Nebraska, USA

REFERENCES

Freedman B (1987) Equipoise and the ethics of clinical research. N Engl J Med 317 (3):141–145
Kincaid H, Dupre J, Wylie A (2007) Value free science: Ideal or illusion? Oxford University Press, Oxford
Schaffner K (1993) Discovery and explanation in medicine. University of Chicago Press, Chicago
Thagard P (2000) How scientists explain disease. Princeton University Press, Princeton

THEODORE M. BENDITT

2. NORMALITY, DISEASE AND ENHANCEMENT*

2.1. INTRODUCTION

The vagueness or imprecision of 'the normal' allows it to be exploited for various purposes and political ends. It is conspicuous in both medicine and athletics; I am going to try to say something about the normal in each of these areas. In medicine the idea of the normal is often deployed in understanding what constitutes disease and hence, as some see it, in determining the role of physicians, in determining what is or ought to be covered by insurance, and in underwriting certain views on social justice. In athletics the use of performance enhancing substances raises different sorts of questions about the normal, especially when it comes to deciding what constitutes an enhancement and which enhancements, if any, are acceptable.

2.2. THE NORMAL IN MEDICINE

There is much writing on what disease is. It is said to be important to understand what disease is because such understanding is thought (often incorrectly, I will maintain) to provide the answer to a number of questions related to health care – namely, the role of medicine and of health professionals, the scope of health insurance, and some of the sorts of help that society owes its members.

In understanding disease, some writers have seen the idea of abnormality as having an important role to play. One line of thought is as follows. Bodily tissue and organs have certain purposes, or functions, or goals (we'll use the word 'function' to mean any of these). Most tissue and organs succeed in performing their functions and when they do, they function normally, or are normal. Whatever fails to perform these functions, then, is abnormal and thus is diseased (or disabled, or disordered, or ...[1]). Ian Hacking, in his book *The Taming of Chance*, credits the 19th century French pathologist Broussais with making this connection; in a representative remark Broussais talks about how something 'can deviate from the normal state and constitute an abnormal or diseased state.' [2]

As more has become known about various organs and tissues, it seems inevitable that the idea of the normal would come to be quantified. Thus, when methods are found of *measuring* the performance of tissue and organs, it can be determined what *range* of activity of, say, a kidney is associated with the functioning of kidneys. This provides a more precise notion of normality, in which abnormality is a matter of being outside the standard or normal range – that is, the range associated with fulfillment of function.

There are two pieces of this approach to focus on: (a) the association of disease with abnormality (indeed, the *identity* of disease and abnormality, as Broussais

H. Kincaid and J. McKitrick (eds.), Establishing Medical Reality, 13–21.
© 2007 *Springer.*

suggests), and (b) the claim that a nonnormative account of disease can be given (and that this is such an account). Most of my comments will be directed toward the first of these (abnormality), so let me first offer some remarks, undoubtedly sketchy, about the second (nonnormativity).

With regard to nonnormativity, the foregoing approach to understanding the idea of disease, which is presented as nonnormative, has been much criticized. The criticisms take one of two forms. One is that the identification of goals and thus of functions is inevitably value-laden; the other says that the idea of *disease* is inherently normative. It seems to me, though, that a nonnormative account of disease is plausible. Consider first the idea that the identification of goals is value-laden: the idea seems to be, for example, that, if the preservation of life is a goal, it is so only because we value it. This idea is confused. Even if it were true that staying alive is *my* goal only if I value it, it does not follow that keeping my organism alive is the goal *of my bodily systems* only if I value staying alive. There are, for example, systems in mammals that maintain body temperature within a certain range, and mechanisms that, when it deviates from that range, operate to get it back within the range.[3] This is a *goal* of that system – identified, I believe, without value judgments – and the system that maintains it has this as a *function* – identified, again, without values. Likewise, there are mechanisms in us that are oriented toward keeping us alive (interestingly, Broussais, according to Hacking, thought there was an *organ* with such a function – he called it the 'stay-alive' organ). For example, if a person falls into deep water there are features of the autonomic nervous system that operate to keep him or her breathing, and being bent on suicide by drowning, and thus *disvaluing* the goal, is not always adequate to overcome the function. Thus, it is *its* function whether or not *we* value it.

It has also been held, alternatively, that the notion of disease is value-laden. So-called normativists about disease hold that whether something qualifies as disease depends on whether it is a *problem* for the person involved. Being farsighted, for example,[4] is a disease (or a disability) only if one is going to spend one's day in the library or the operating room, and a failure of a major organ such as a kidney is not a disease 'if other opportunities exist for performing the same function.'[5] Whether something is a disease, then, ultimately depends on whether the owner of the organ *wants* it to do what those organs usually do, so it's a subjective and/or value-laden matter. As this view has a lot in common with what follows, I will postpone comment on it.

With regard to the association of disease with abnormality, there are many who have substantial doubts about the purported connection, at least regarding the idea that abnormality is, or is sufficient for, disease. This line of thought has been developed, with impressive examples, by Alice Dreger and by Anita Silvers. A serious difficulty with understanding disease in terms of abnormality, as they see it, is that it turns whatever is *different* in its functioning into disease or disability. Furthermore, in so doing it underwrites a policy of dealing with these situations by attempting to restore or approximate normal function even if that is not best for the individual involved. The idea of disease-as-difference, they maintain, leads, in a useful phrase, to the *medicalization of difference*. Dreger, for example,[6] decries the assumption that psychosocial problems resulting from stereotypes (regarding, for example, short stature or mixed genitalia) should inevitably be 'fixed' by 'normalizing' the anatomy. Some conjoined twins, she observes, are better off, particularly in terms of their identities and individuality, remaining conjoined, as opposed to trying to separate and fit into others' conceptions of these things. Silvers too is concerned about the attempt to force on people the restoration of species-typical modes of functioning.[7] She observes that there has historically been a preference in schools for the deaf for oralism rather than signing – 'When oralism dominated in schools for the deaf, deaf children could either try to lip read and speak, or have no education at all.' And: 'for Canadian children with no usable lower limbs, mechanical limbs were the only mobility option offered.' That is, a 'natural' form of mobility (walking with braces) was preferred to an 'unnatural' one like using a wheelchair, even if the latter would be better for the person involved. What Dreger and Silvers highlight is that the medical (restore-the-normal) approach seems to hold that lesser capacity in a wider social environment is preferable to more limited, but more effective, capacity in a limited environment. And further, they observe that the idea that restoration of normal function would be beneficial presumes that the social environment is held constant, whereas some people might prefer to try to alter the social environment rather than alter themselves to try to fit into the existing environment.

There is undoubtedly truth in the examples presented above, but it may be misdirected. The problem may not be (or at least, may not always be) in seeing disease as abnormality. The problem may be rather with the idea that calling something a disease automatically puts it into the category of 'something that needs fixing'. Putting it somewhat differently, the difficulty is in insisting that there is a tight connection ('tight' as in 'definitional' or 'entailment') between the idea that something is a disease and that it ought to be fixed (or even 'prima facie ought to be fixed' or 'ought to be fixed, other things being equal'). It is hard to accept, for example, the idea that a nonfunctioning kidney does not qualify as diseased simply because an artificial kidney would be preferable. Those who argue in this way see the notion of the normal as the culprit – the normal is seen as the good, the desirable, the best, and deviation from it calls for its reinstatement, if possible, or as close to it as can be achieved. But the culprit is not the normal; the problem is not with understanding disease in terms of abnormality; rather, it is, I believe, the presumed tight connection between 'is a disease' and 'ought to be fixed'.

A good way to undermine this presumed connection is to alter the language we use to represent the situation. Let's use the word 'broken' instead of 'diseased'. Instead of saying that a nonfunctioning kidney is diseased, let's say simply that it is broken. Now it seems to me that saying that something is broken does not carry the implication that it ought to be fixed – we do not say that something is broken only if we think it ought to be fixed. Likewise, thinking that it ought not to have anything done to it[8] in no way suggests that it is not broken. My computer, my dishes, can be broken whether or not they ought to be fixed. If I find some broken parts of an antique doll in an old box in my basement, I might or might not want to get the item fixed, though I would hardly think 'Well, it's broken, so it is certainly to be fixed – unless I have a good reason not to.' Further, if I finally decide against doing anything to it, I surely will not entertain the idea that it is not broken. Similarly, it seems to me, it's misleading to say that a kidney that can no longer cleanse blood is indicative of disease only if the owner wants it to cleanse blood. It *is* broken – or, if you like, diseased – whether or not the owner wants it to do that. On the other hand, saying that it is broken does not mean or imply that it ought, even prima facie, to be fixed. The question of what the involved person wants is relevant not to whether there is disease, but to what is to be done about it. So far, then, the nonnormativist account of disease holds up (or so it seems to me).[9]

This is not to say, though, that there are no problems in the use of the notion of normality. Nor, in particular, is it appropriate to come to the conclusion that wherever there is abnormality there is disease – let alone (as Dreger and Silvers argue) that there is *ipso facto* a role for medicine. The idea of the normal, as in 'normal function', makes the most sense where there is something, such as a piece of tissue or an organ, whose function can be ascertained without value judgments and is carried out without one having to *do* anything specifically related to functioning of that tissue or organ. In such cases, I have maintained, it is appropriate to identify abnormality with disease. The idea is readily extended to certain situations in which there is abnormality in outward movement or behavior. For example, the legs, or whatever controls their function, can be damaged, which shows up in the attempt to walk. In this case the behavior is abnormal, though it is brought about by another, underlying, abnormality that constitutes the disease or disability.

The next extension, and a troubling one, occurs, however, when it is *behavior alone* that is abnormal. Certainly where there is an organ that does not fulfill its function there is a case for identifying abnormality with disease (or so I have argued). But is this so in the extended case, with the idea of the abnormal unmoored from a specific function? When we are talking purely about behavior, but not functionality, is there a case for saying that difference is equivalent to defect? Or is there only difference?

This is an issue that has been much discussed, and is readily recognized, in the area of *mental* illness. Thomas Szasz's views on this topic are well known. He maintains, in effect, that there is only difference – that is, that behavioral abnormality by itself does not imply disease. Szasz insists, along the lines sketched here, that in 'real' illness there is always an underlying pathology that causes the outward occurrences that are seen as problematic,[10] but that absent an underlying pathology there are only 'problems in living' – i.e., difference. Szasz was arguing against the engrained, largely Freudian, view of the psychiatric profession that abnormal behaviors *constitute* mental illness. In rejecting this, though, Szasz seems to have gone to the opposite pole, holding that there is no mental illness at all, that no so-called mental illnesses qualify as disease in the proper sense. This extreme position is now largely rejected. Allan Horwitz, for example, following Jerome Wakefield's account of mental disorder as involving dysfunction, maintains that there are indeed genuine mental disorders, such as psychotic disorders and bipolar disorders.[11] (On the other hand, it is robustly maintained by some that there are pathologies, such as depression, that do not necessarily involve the malfunction of anything – that may indeed involve the proper functioning of internal mental mechanisms in environments in which proper function is experienced as unwelcome. It's a good question, about which I venture nothing, whether such phenomena should be regarded as diseases.)

In the foregoing part of the discussion, I have maintained several points. I have argued that where some tissue or organ is abnormal in failing to perform its function, it makes sense to talk of disease and that this is not a normative or value-laden notion of disease. I have argued that where there is difference alone (abnormality independent of failure of function) there is not (at least so far) a case for talking of disease or disability. And third, I have argued that in any case the fact that something is a disease, let alone its being simply an abnormality, does not by itself imply the appropriateness of medicalization. In addition, though I have not discussed it, it seems to me that there is no reason, despite anything said above, to take it that the role of physicians, or of medicine, is limited to dealing with disease – there are undoubtedly other sorts of things for which the skills of physicians can be called upon, in which medicine can provide relief. 'Disorder status and appropriateness for treatment are clearly two different concepts ... .'[12]

2.3. THE NORMAL IN ATHLETICS

As we have seen, the idea of the abnormal, according to some, plays an important role in defining the role of medicine. There is, though, another side to this – that the idea of the normal plays a role in defining the *limits* of medicine. On the one hand, it is often said, medicine is concerned with disease, which may be understood in terms of deviation from the normal, which it is the role of medicine to restore. But, the line of thought frequently continues, it is not the role of medicine to go beyond this, to provide enhancements. For some this is a conceptual matter: Medicine has always been about healing. The role of physicians is not a matter of what they may have the knowledge and skills to do, but about what constitutes healing. Enhancing, though, is not healing. Therefore it is not within the purview of medicine.

For others, limiting medicine to healing is a political matter governed by the idea of equality of opportunity.[13] Equal opportunity is a matter of people being in a position to make the most (socially, economically) of what they have in them, their natural endowments. A just society tries to remove barriers to the individual's achieving this, though owing to differences of endowment, people realize different economic and social benefits. Enhancements, on the other hand, are problematic because they alter the allocation of abilities and thus upset the outcomes that equal opportunity is intended to achieve.

Within this framework there are difficult questions, about what constitutes an enhancement that I will mention but not try to answer. For example, is the pharmacological improvement of a person who is below average (or perhaps *well* below average) an enhancement? Or is an improvement an enhancement only if the person is already at least average? Is a strengthening of a person's immune system, such as by a vaccine, that doesn't enable the person to do something better but only *prevents* her getting ill, an enhancement?[14] Is improving the memory of a 75-year old, to make it as good as when she was much younger, an enhancement, or is it only the restoration of a function? And how shall we think about a medical response to a deficiency that has the result not only of getting rid of someone's problem but making him superior to others as a by-product. It has been observed, for example, that this can be an outcome of the use of Ritalin in dealing with attention deficit disorder.

The topic of enhancement is not limited, however, to defining medicine or to the political issue of equality of opportunity. There is also a great deal of concern about enhancement in athletics, and what I want to suggest in what follows is that the idea of the normal – as in, the normal competitor – does play a kind of limiting role in this area.

Let's begin by asking what's wrong with certain enhancements, such as the use of steroids, in athletics. One answer is that they can be dangerous, as has often been demonstrated. A related matter is that this danger tends to spread as knowledge of the potential benefits of steroids spreads. This is said to be of concern in two ways. First, there is a coercive element: others feel compelled to use them simply to be competitive. Second, there is an impact on younger people who aspire to athletic prowess, owing to the phenomenon of athletes as role models.

A second answer is that use of performance-enhancing drugs gives users an unfair advantage, as countless weight-lifting, sprinting, and other competitions have shown. Numerous record-holders, as it turns out, have used these drugs. It has been observed, in this connection, that there is a potential irony regarding performance enhancement, for if all competitors use these substances the advantage is neutralized – nobody is better off than if no one used them. (Perhaps it should be noted, though, that this is true only among current competitors; everyone who uses steroids has an advantage vis-à-vis past competitors – in the setting of records.)

Danger and fairness are certainly issues with respect to enhancement, but they are not the only issues. There is something else that is problematic in performance enhancement. It has to do with what we think competition is about in the first place.

We begin with the knowledge that humans are different – we have different abilities. One of the major questions for human beings has always been what to do about this, to which well known political theories give answers. But there has, always, I suppose, been a different sort of response as well: to compare, to see who is best at various things, to determine who is fastest, who is strongest, who is smartest, who is most attractive, who is most skilled, etc. In some cases we want to know in addition what the limits of human capability are – what is the most that a human can lift, or what is the fastest a human can run, and so forth. Competitions are good ways to try to discover these things. In what follows I am going to use sprinting, foot racing, as a standard example.

If we want to determine who is the fastest, we certainly want the race to be fair. There are, though, different ways in which a race can be fair or unfair, only some of which are relevant to determining who is the fastest. A race can be unfair if one competitor starts ahead of another. This is an unfairness that affects the determination of who is fastest, so it must certainly be eliminated if we are to determine who is fastest. It is important to be clear about the way in which this unfairness affects the determination of who is fastest. The point is that, since the competitors are not running the same distance, it is impossible to determine, in that competition, who has traversed the distance in the shortest time. On the other hand, if a competitor takes a drug that enables him or her to run faster, then he actually *does run faster* over the prescribed distance. *If* there is unfairness in this, it is in the drug user having an improper advantage over other competitors; it is not an intrinsic unfairness in the race itself.

Well, is it unfair, or wrong in some other way, for a competitor to use a performance-enhancing drug? (That is, is it unfair apart from its being prohibited by rules? For if there are rules against it, it is certainly unfair.) That depends on why we want to have races at all. It depends on what we want to find out by means of a race.

Suppose a competitor shows up for a race with bionic muscles, or with wheels implanted into his feet. No doubt there is something wrong with this. But what is it? Is it the unfairness to the other competitors? No doubt it is unfair, but this is hardly the only thing wrong with it, for it seems clear that there would still be something wrong even if other competitors, or all of them, showed up with similar implants. The problem, it seems to me, is that the enhancements in question are incompatible with the point of the race. The point was not merely to determine a winner in a fair race. The point was to determine a winner in a race of a certain kind, and wheel implants make it a race of a different kind.

What kind of race are we interested in? What is the point of the races we conduct? Let's begin by identifying a few obvious points. First, the point of a race is not merely to determine a winner. If it were, it wouldn't matter who the competitors were. We do not, for example, race children against adults or men against women or humans against dogs or horses. (We don't *normally* do this; sometimes we do, but for off-beat reasons – for the spectacle, or if we think the child or the woman can win, or the horse or dog lose.) Second, the point of a race is not (or at least not usually) merely to see how fast a human can traverse a given distance. For if that were all we cared about, there would be no problem about bionic muscles or wheel implants. (An analogy to an auto race might be useful. Suppose one of the cars showed up with a rocket attached. Of course this is unfair to the other racers, but, unless we're interested solely in how fast a car can traverse a given distance, it's not the only concern. What we are interested in is how fast a car *of a certain sort* can do it.) Of course, we *might* be interested in these things. We might be interested in technological innovations, or we might really be interested in how fast a human or a car can *in any way at all* traverse a given distance (as indeed, there are competitions among rocket-assisted cars to try to establish a land speed record).

But in races among humans we are not usually interested in these things. We are usually interested in how fast an *unaided* human can do it.

This is, I believe, where the notion of the normal enters, both for cars and for humans. The normal when it comes to foot races defines the class of things among which a race would be of interest. The idea of the normal human being, for these purposes, is a shorthand way of referring to *how unaided* a human must be to make the result of the competition something we are interested in. Having bionic legs, for example, puts a person outside the class of normal humans for the purpose of the sort of race we are interested in, quite apart from the issue of fairness to other competitors.

The phenomenon of chess matches between human and computer is a useful analogy. These are interesting in their own right, but I doubt we will ever see tournaments, except as novelties, with both human and computer entrants. We will have competitions among humans, and competitions among computers (or among the humans who program them). But now what would we think if a chess player were to come along with a chip implant that enlarges his or her memory and thus the capacity to analyze moves? We are likely to want to have competitions among 'normal' chess players, and perhaps specialty competitions for enhanced ones. Of course, if we ever reached a point in which chip implantation became standard for humans (infants might get their chip implant along with their shot of vitamin K), that would *redefine* what we call a normal human and thus a normal chess player. This would then constitute the group among whom such competitions are interesting.

In sum, while I am doubtful that the idea of the normal defines limits for medicine or for physicians (though it might form the basis for limits in insurance or for certain social policies), it seems to me that it *is* a properly limiting notion in other areas of life.

Department of Philosophy, University of Alabama at Birmingham, Alabama, USA

NOTES

* This is a slightly revised version of a talk given at the Conference on Philosophical Issues in the Biomedical Sciences, sponsored by UAB Center for Ethics and Values in the Sciences, Birmingham, AL, May 14, 2004. I would like to thank Christopher Boorse for some helpful comments.
[1] For purposes of this discussion, I am treating the various terms that are used, such as disease, disability, disorder, as if they mean the same. This is a simplification. Ron Amundson, in 'Against Normal Function,' in *Studies in History and Philosophy of Science Part C: Studies in History and Philosophy of Biological and Medical Sciences*, 31,1 (March, 2000, pp. 33–53), distinguishes between 'permanent and stable conditions, commonly called disabilities' and 'the more episodic or life-threatening conditions commonly called diseases,' and indicates a primary interest in the former, whereas this paper focuses more on the latter. It may be that discussions of the role of the abnormal in medicine do not pay enough attention to differences between disease and disability.
[2] Ian Hacking, *The Taming of Chance* (New York: Cambridge University Press, 1990), pp. 82, 165.
[3] See generally Christopher Boorse, 'A Rebuttal on Functions,' in *Functions: New Essays in the Philosophy of Psychology and Biology*, ed. by André Ariew, Robert Cummins, and Mark Perlman (New York: Oxford University Press, 2002), pp. 63–112, at 68ff.

[4] These examples come from Arthur Caplan's explanation of this approach. See Arthur L. Caplan, 'If Gene Therapy is the Cure, What is the Disease?' in *Gene Mapping: Using Law and Ethics as Guides*, ed. by George J. Annas and Sherman Elias (New York: Oxford University Press, 1992), pp. 128–141, at p. 132.

[5] Ibid., p. 133.

[6] See Alice Dreger, 'When Medicine Goes Too Far in the Pursuit of Normality,' *New York Times*, July 28, 1998.

[7] Anita Silvers, 'A Fatal Attraction to Normalizing: Treating Disabilities as Deviations from 'Species-Typical' Functioning,' in *Enhancing Human Traits: Ethical and Social Implications*, ed. by Erik Parens (Washington, DC: Georgetown University Press, 1998), pp. 95–123. Quotes appear at p. 114.

[8] Originally the text read 'thinking that it ought not to be fixed ...', but a reviewer pointed out that the term 'fixed' already implies that the thing is broken.

[9] Fred Dretske argues that the term 'broken' *is* normative, for to say that a thing is broken is to say something about how it is supposed to be. If I were to make a doll (or doll parts) identical to the broken doll, it would *not be broken*. This shows that something can qualify as broken only if it has a history such that it 'departs in some degree from a standard [perhaps derived from the intentions of a designer or maker] that defines how it should be.' Hence, 'broken' is normative. This seems right, but it is not the right sort of normativity for those who are normativists about disease. That the correct application of a term involves a normative judgment does not mean that it is normative in the sense of implying anything about how things *should* be (in the future). Specifically, the norms involved in applying the term 'broken' do not imply that a broken item must or should be repaired. However much an object departs from what it was designed to be, such that we can correctly call it broken, that does not mean it should continue to be the way it was designed to be, and therefore its being broken does not imply that it should be repaired. Fred Dretske, 'Norms, History and the Mental,' in *Naturalism, Evolution and Mind*, ed. by Denis M. Walsh (Cambridge, UK: Cambridge University Press, 2001), pp. 87–102, at p. 90.

[10] See Thomas Szasz, *Insanity: The Idea and its Consequences* (New York: J. Wiley, 1987), p. 12, and *Cruel Compassion* (Syracuse, NY: Syracuse University Press, 1998), p. 162.

[11] Allan V. Horwitz, *Creating Mental Illness* (Chicago, IL: University of Chicago Press, 2002).

[12] See Jerome C. Wakefield, 'Spandrels, Vestigial Organs, and Such: Reply to Murphy and Woolfolk's 'The Harmful Dysfunction Analysis of Mental Disorder',' *Philosophy, Psychiatry, & Psychology*, 7.4 (2000), pp. 253–269, at p. 258.

[13] See Norman Daniels, 'Justice and Health Care,' in *Health Care Ethics*, ed. by D. Van DeVeer and T. Regan (Philadelphia: Temple University Press, 1987), and D.W. Brock, *Life and Death* (Cambridge, UK: Cambridge University Press, 1993).

[14] The issue might be raised in the following way: Is it sufficient for an enhancement that one's 'equipment' is improved, or must one also be enabled thereby to carry out certain specific functions better? Improving the immune system does the former, but not the latter.

3. HOLISTIC THEORIES OF HEALTH AS APPLICABLE TO NON-HUMAN LIVING BEINGS

3.1. INTRODUCTION

This paper will deal with the following question: Is a so-called biostatistical theory of health and disease more universal than a holistic one in the sense that it is applicable not only to human beings but also to other animals and plants? I will take as my starting-point the distinction between the two types of theory. The biostatistical theories characterize health and disease in biological and statistical terms, such as biological functioning, survival, and normality (Boorse, 1997, pp. 1–134; Schramme, 2000). The holistic theories tend to use psychological or sociological terms, such as terms denoting feelings, actions, goals, and sometimes societies (Nordenfelt, 1995; Fulford, 1989). All these theories are primarily intended to be theories about human health and disease. In particular the holists have introduced notions and arguments that are particularly applicable to human conditions. On the other hand, our ordinary notions of health and disease are indeed also used with regard to the worlds of animals and plants. It is also obvious that our ordinary uses of the terms 'health' and 'disease' are similar with regard to all living beings.

From this point of view it may be seen as an advantage of a biostatistical theory that it easily applies to the domain of animals and plants. The crucial notions in such a theory, viz. survival of the individual and of the species, are applicable to all living beings. However, it seems doubtful, whether holistic notions such as feelings, actions and intended goals can be attributed to animals and, in particular, plants. Should this, then, be regarded as a general weakness in the holistic theories of health and disease?

This issue will be the subject of my paper. I will try to demonstrate that the holistic approach is equally valid for the analysis of health when it comes to animals and plants. This does not mean that I think that exactly the same holistic concepts are applicable to all living beings. However, I suggest that there is a core holistic foundation common to the notions of health and disease in all cases. I will argue for this thesis in some detail for the case of animals and I will at least suggest how a similar argument can be performed with regard to plants. As one of the platforms for my analysis I will use my own theory of health as presented in *On the Nature of Health* (1995) and *Health, Science, and Ordinary Language* (2001).

The starting-point for my analysis will be the biostatistical theory of health and disease developed by Christopher Boorse during the 1970s and revised in his famous paper from 1997: 'A Rebuttal on Health'. There are two central definitions that form the basis of Boorse's characterization of health. There is first the definition

H. Kincaid and J. McKitrick (eds.), Establishing Medical Reality, 23–34.
© 2007 *Springer.*

of disease: 'A *disease* is a type of internal state which is either an impairment of normal functional ability, i.e. a reduction of one or more functional abilities below typical efficiency, or a limitation on functional ability caused by environmental agents.' Second, there is the definition of health based on this characterization that says laconically: Health is identical with the absence of disease.

The crucial concept here is functional ability. This has been explained at other places in Boorse's presentations. An organ exercises its function, for instance the heart is pumping in the appropriate way, when it makes its species-typical contribution to the individual's survival and reproduction. Survival and reproduction are the crucial biological goals, according to Boorse. The notion of biological function is tied to these goals.

Boorse's elegant theory seems to have several advantages. The notions of survival of the individual and survival of the species fall well within evolutionary theory, which is much celebrated in both biological and sociological contexts these days. It is true that Boorse does not explicitly found his ideas on evolutionary theory in his writings but roughly the same ideas have been taken up by Jerome Wakefield (1992 and later), who very clearly puts them into the evolutionary framework in his explications.

Another advantage of Boorse's biostatistical theory, as I have indicated, is that it can be quite easily generalized and applied to other living beings than humans. Survival of the individual and survival of the species are notions applicable to all living beings. Not only apes, but also worms and amoebas can die and fail to reproduce. And the same holds for all plants, from orchids to mosses.

In contrast to Boorse's biostatistical theory we have a number of holistic theories that do not accept Boorse's criteria of health and illness. At least they do not accept these criteria as sufficient for the analysis of health and illness. It is significant that the holistic theories refer not only to the survival but also to the *quality* (mainly the welfare) of the life of the individual. According to these theories, a person can be ill, not only if the probability of the person's survival has been lowered but also if he or she does not feel well or if he or she has become disabled in relation to some goal other than survival. In his classical analysis of health Galen, from the first century AD, says that 'health is a state in which we neither suffer pain nor are hindered in the functions of daily life' (translated by Temkin (1963), p. 637). K.W.M. Fulford (1989) says that 'people who are ill are unable to do the things that people ordinarily just get on and do, moving their arms and legs, remembering things, finding their ways about familiar places and so on.' In my own characterization, which primarily concerns the notion of positive health, I say that a person is completely healthy if, and only if, this person is in such a state that he or she can realize his or her vital goals given standard circumstances. My notion of a vital goal is technical: A vital goal of a person is a state of affairs that is such that it is a necessary condition for the person's minimal happiness in the long run. For further clarification see Nordenfelt, 1995, p. 93.

I have elsewhere commented on the Boorsian analysis of health and disease with regard to human beings from different points of view. Here I will concentrate only on an argument, which I think has the greatest force in the present discussion. This is an argument that takes as its starting-point the origin of the human language of illness and health. The idea here is that holistic *illness* is the primary notion and that disease and health are secondary notions in this derivation.

3.2. THE REVERSE THEORY OF DISEASE AND ILLNESS

I wish first to argue that the notion of illness when it comes to humans has its basis in the existence of a perceived *problem*. I therefore endorse the *Reverse Theory of Disease and Illness*, which was first suggested by Georges Canguilhem in 1943 (first published in English 1978), and later developed by K.W.M. Fulford in 1989. Let me introduce this idea with the help of the following, hopefully plausible, story with regard to the emergence of the concepts of illness and disease.

1. In the beginning there were people who experienced problems in and with themselves. They felt pain and fatigue and they found themselves unable to do what they could normally do. They experienced what we now call illnesses, which they located somewhere in their bodies and minds. Several people came to experience similar illnesses. This led to the giving of names to the illnesses, and hereby the presence of the illnesses could be efficiently communicated. This was the phase of *illness recognition* and *illness communication*.

2. The people who were ill approached experts, called doctors, in order to get help. They communicated their experiences to the doctors, via the illness language. The doctors tried to help them and cure them. In the search for curative remedies, the doctors did not just rely on the stories told by the people who were ill. They also looked for the *causes* of the illnesses within the bodies and minds of the ill. This meant in the end that they initiated systematic studies of the biology of their patients. This was the phase of *search for the causes of illnesses.*

3. As a result of these studies the doctors found some regular connections between certain bodily states and the symptoms of their patients. They formed hypotheses about causal connections between the internal states and processes and the illness-syndromes. They designated these causes of illnesses *diseases*. And they invented a vocabulary and a conceptual apparatus for the diseases. This was the phase of *disease recognition.*

This is a quasi-historical sketch of the development of the notion of disease. According to this story the concept of illness is primary to the concept of disease. At the heart of the story lies a problem that has to be solved, through an investigation into the causes of this problem. These causes are assumed to exist within the subject's body or mind.

I find this sketch to be a plausible one also for the explication of the contemporary concepts of illness, disease, and health. Diseases are typical causes of such problems as we call illnesses and for which we also today approach doctors. This explication is more plausible than the rival one that says that diseases are bodily states that make a statistically subnormal contribution to the survival and reproduction of their bearers. A problem constituting illness need not entail a threat to or a reduced support of the person's life or reproduction. The problem quite often concerns pain, other kinds of suffering, or disability. And the subject normally correctly believes that this problem has some kind of internal (biological or psychological) cause. I therefore conclude that the concept of human disease is related primarily to suffering and disability and not to the increased probability of death.

From the concept of illness we can, if we wish, negatively derive the concept of health. Health is the absence of illness (the holistic concept of illness) and not the absence of disease, as Boorse describes the situation. However, if we wish to define optimal health we must do it, I think, in the positive way I indicated above.

The primary focus of attention is thus the illness – the problem as perceived normally by the subject. From the concept of illness we can derive the concept of disease, i.e. the internal state which causes (or tends to cause) the illness. But observe here how the diseases are identified. They are identified on the basis of an illness-recognition. A discovery of the disease presupposes the occurrence of an illness. Hence, the expression 'reverse theory of disease and illness'.

Given this interpretation, we arrive at a definition of disease, which differs, from the Boorsian biostatistical one. A preliminary definition of disease would thus be: Disease = a bodily or mental process, which is such that it tends to cause an illness (understood as a state of suffering or disability experienced by the subject). Certainly, many of the conditions picked out by the Boorsian criteria will be identical with the ones picked out by the holistic criteria. A cancer is a disease for Boorse as well as for myself. But the reasons differ. For Boorse a cancer is a disease because it makes a statistically subnormal contribution to the subject's survival. For myself, however, a cancer is a disease because it tends to cause its bearer disability and suffering. But, and this is important, some diseases picked out by the holistic criteria will not be counted as diseases by the Boorsian ones. A person may be in pain and disabled by internal bodily causes without this condition lowering the probability of the person's survival. But the converse may also hold, there may be diseases picked out by the biostatistical criteria, which are not picked out by the holistic ones.

Suffering and disability are not novel concepts in the history of medicine or health care. They are central in the clinic but they have not been consistently used there in the theoretical characterization of basic medical concepts such as disease and pathology. In the history of philosophy of health, however, we can find the ideas of suffering and disability in the writings of some authors. I have already quoted Galen in his *Techné Iatriché* (*The Art of Medicine*), written in about AD 190. (For a modern source, see Galen, 1997).

There are alternative ways of using the crucial concepts of suffering and disability in the construction of theories. Either one uses both kinds of concepts, as Galen does and as Lawrie Reznek (1987) does, and says that illness is constituted by both suffering and disability, or by either suffering or disability, or one focuses on one of them for the purpose of definition. A number of contemporary theorists, myself included, have focused on the pair of concepts ability and disability since we find it to be more universally useful than the pair well being and suffering.[1] My subsequent reasoning will provide further arguments for an ability approach to health and illness.

However, for the moment I will disregard these differences. I think *all holists* could get together in maintaining that illness has its conceptual root in the notion of a *problem* with one's body and mind, and that, conversely, health is a state where there are no such problems and where one's life has a certain high quality. I will therefore not have my own proposal as an exclusive platform.

3.3. INTRODUCING A HOLISTIC VIEW OF ILLNESS AND HEALTH INTO ANIMAL SCIENCE

Is it possible, then, to apply an analogous reasoning in the animal context? Can we talk about illness, disease, and health in a holistic sense with regard to animals?

First, a word of caution. What animals are we talking about? Ultimately, I wish to extend my argument to all animals and even try it on plants. However, in a short paper like this I can only exemplify from a very small number of animals. Moreover, most of these examples will concern animals, like dogs and horses, which are close to humans. Hopefully, however, I will be able to show how the argument can be extended to cover other animals, further down in the developmental hierarchy of animals.

A first crucial difference between humans and animals is the following: There is in the case of animals no subject that can consciously present a problem using an illness language like the human one. Moreover, there is no animal subject approaching a physician and explaining what the problem is. Can the holistic concept of illness, then, be the core concept in the animal context?

Let me try to transpose my argument to the animal world. It is true that most animals (as far as we understand) do not embrace a full-fledged language. On the other hand, most other elements present in my human story can be present in a parallel animal story. Most animals can have problems. Many of them, I wish to argue, can suffer and can express their suffering. In their wordless way they can ask for help. If the animals in question are in close contact with humans, which is the case with pets and livestock, the humans can interpret the call for help and can try to respond to it. If the humans suspect serious illness, they call for further support and approach a veterinary surgeon, who will act very much like the physician in searching for an underlying disease responsible for the illness of the animal.

I have argued with regard to the human case that it is unreasonable to define disease or pathology simply in terms of malfunction in relation to survival and reproduction. I think the case is similar on the animal side. Animals, like humans, can be ill without there being any threat to survival and reproduction. They can be ill in the sense that they feel malaise and have a reduced capacity, as humans can. We can easily observe when our pet dog is not feeling well at all, when it drags its tail or even whines from pain. A horse can have a disease that reduces its capacity to run in a race. This is a reduction in relation to a goal, but this goal is not identical with the survival of the horse.

But can we make this move as easily as this? Can we substantiate the claim that animals have feelings? Can animals suffer? This is a complex story. There are indeed a few animal scientists who doubt the possibility of animal feelings altogether (Barnard and Hurst, 1996). On the other hand, the overwhelming majority of veterinarians and animal scientists base their whole work on the presupposition that many animals can feel pain, nausea, and fatigue in much the same way as humans. Why should there otherwise be so much concern about animal welfare (Dawkins, 1990; Duncan, 1996)?

But can animals feel anything but simple sensations? Is there any point in introducing the more complex vocabulary of feelings? Common sense certainly points in this direction. According to common sense my dog can clearly show deep affection to my family and myself. This is an emotion of love or friendship. My dog can show a high degree of expectation and hope when I move toward the door and thereby indicate that I may be going out for a walk. And certainly my dog can be sad and depressed about the fact that we are leaving home and abandoning him for a while.

Moreover, specialists in animal science substantially support these commonsense conceptions of the diversity of feelings of animals. Donald Broom (1998) has systematically characterized a variety of complex animal feelings, including both emotions and moods. Broom discusses in some detail the emotions of fear, grief, frustration, and guilt in animals. In the case of fear Broom uses the following definition: 'a normal emotional response to consciously recognized external sources of danger' (p. 383). He notes that this is an extremely common emotion in vertebrates, one that has an obvious role in the animal's coping process. Concerning grief, Broom notes that there are many reports of pets, especially dogs and monkeys, that show the same sort of behavior that humans in such circumstances show. 'This behaviour is sometimes described as mourning and has also been described in horses, pigs and elephants' (p. 385). With regard to cognitions Donald Griffin has in his celebrated *Animal Thinking* (1984) argued strongly for the existence of complicated thinking among many animals.

One may seriously doubt, however, that all animals have cognitions, sensations and emotions. What about worms, cuttlefish, and indeed amoebas? Although I think that they must all have some minimal perception (i.e. some mechanism to get information from the outside), it is probably true that the most primitive ones lack most other mental properties. For one thing they have only a rudimentary neurological system or, in the case of amoebas, none at all. But such an observation is a problem only to those holistic theorists who base their idea of illness *totally* on suffering. Higher animals may suffer, but not all animals do. A theory of health and illness regarding these lower-level animals cannot depend, then, on the notion of suffering.

What, then, about the ability theories of illness and health? Can they be of further help? Consider, for instance, my own suggestion that health is constituted by the person's ability to realize vital goals. Crucial such goals in the human case are related to the subject's daily living, to his or her occupation, and to his or her close human relations. Most of these goals are peculiar to humans.

However, this ability theory of health does not concretely specify the vital goals. The animals need not have exactly the same vital goals as humans. Animal science clearly shows that animals have goals. Many animal theorists claim that the model of animals as goal-seeking systems (Toates, 1987) has largely replaced the model of animals as stimulus-response automata. And researchers such as Stamp Dawkins (1990) and Ian Duncan (1996) have devised sophisticated experiments putting animals in situations of choice in order to determine the goals of the animals. They have demonstrated that many animals can make informed choices based on earlier experiences. We can, they say, ascribe a want-faculty to most animals. A predator wants to get its prey; the predated animal wants to escape. All animals (that reproduce in a sexual way) want to mate and have offspring, etc. But the animal can also relate to its own sensations, emotions, and moods. The cat that has an itch in a leg wants to get rid of it, and therefore scratches the leg. The animal in pain focuses its attention on the pain and wants to get rid of it.

Moreover, it has also been shown that low-level animals with a limited neurological make-up can learn from experience. There are recent remarkable findings concerning fish and invertebrates with regard to the faculties of cognition and learning (Bennett, 1996, pp. 219–224; Karavanich and Atema, 1998, pp. 1553–1560; Bshary et al., 2002, pp. 1–13).

Hence, not only humans perform intentional actions, although we hardly ever use the term 'intentional action' in the animal case. However, we would not understand the 'doings' of many animals unless we were to use our own mental language and the associated action language. So, ascribing intentional actions to many animals is almost unavoidable. Therefore, there is nothing incoherent in ascribing goals to them. But what about *vital goals* in my technical sense? Neither a dog nor indeed a worm embraces the concept of a vital goal. No, but neither does a baby or a senile person, nor for that matter most other people. The notion of a vital goal is a theoretical notion; it is not tied to any hierarchy of preferences in a psychological sense (although I contend that most vital goals are also strongly preferred in a psychological sense). A goal is vital to an animal, according to my theory, as it stands if, and only if, its realization is a necessary condition for the animal's long-term happiness. And there is nothing unrealistic in using these concepts with reference to a large part of the animal world.

But what is included among the vital goals of animals? Again, we can find many analogies to the human situation. There are certainly activities of daily living associated with all animals. The animals take care of their hygiene; most of them build their own nests, lairs, etc. and at least the wild ones are quite efficient in feeding themselves. If they were to be prevented from performing these tasks they would suffer.

But again of course a problem crops up. I may have argued convincingly for the case that the *higher*-level animals can have vital goals in my technical sense. But what about the lower-level animals? We may perhaps say about all of them that they have goals and that they pursue goal-directed 'doings'. But how can they have *vital* goals in my sense, since such goals are, per definition, related to long-term happiness? How do we characterize the happiness of a worm or an amoeba?

My answer to this is that we need for the present purpose a more general notion than that of happiness. The notion I propose is that of *welfare*. And in order to substantiate the idea that lower animals can have welfare, let me turn to the philosophy of nature presented by Paul Taylor (1986).

Taylor presents what he calls the biocentric outlook in order to develop an environmental ethics that is not dependent on human interests. This outlook contains the following elements:

1. From the perspective of the biocentric outlook one sees one's membership in the Earth's community of Life as providing a common bond with all the different species of animals and plants that have evolved over the ages.

2. The biocentric outlook on nature also includes a certain way of perceiving and understanding each individual organism. Each is seen to be a teleological center of life, pursuing its own good in its own unique way. Consciousness may not be present at all, and even when it is present the organism need not be thought of as intentionally taking steps to achieve goals it sets for itself. Rather, a living being is conceived as a unified system of organized activity, the constant tendency of which is to preserve its existence by protecting and promoting its well being[2] (A summary from pp. 156–158).

In my brief discussion about feelings above I admitted that we have little evidence for talking about feelings in the case of animals with rudimentary or nonexistent neurological make-ups. Here I also think that the holistic characterization of health must use a more generic concept. When the goal of a worm is frustrated, then the worm, as far as we know, does not feel pain. However, something negative has happened to this worm. The welfare of the worm or the quality of the worm's life has been reduced. We can infer this from certain behaviors or non-behaviors on the part of the worm. It may make stereotyped and unsuccessful movements, for instance with regard to reaching a goal that it has started to try to reach. Or it may not move at all. We may also compare this worm with another worm creeping just beside it. The latter worm is lively, it moves around quickly, it is bigger and it seems to be thriving. The first worm, we say, must have some problem. It is ill.

I now also tentatively approach the universe of plants and the matter of how we might extend a holistic approach to the understanding of the health and illness of plants. The notion of happiness is clearly not applicable to plants. On the other hand we may be able to talk about a reduced quality of the plant. We can compare stages in the life of a particular plant. One year it flourishes and grows a lot. Another year it shrinks and even withers and does not produce any flowers. A third year the plant is revitalized and is flourishing. When it flourishes the quality is high, and the plant is well; when it withers the quality is low and the plant is ill.

My suggestion for the systematic characterization of health and illness in the case of animals and plants is then the following: An animal A or a plant P is healthy if, and only if, A or P has the ability to realize all its vital goals given standard circumstances. A vital goal of A's or P's is a necessary condition for the long-term welfare of A or P. In the case of the higher animals the criterion of welfare is the happiness of the animal. In the case of lower-level animals and plants we must find other criteria, such as the vitality of the animal or plant. Observe also that in the case of plants we cannot talk about actions or 'doings'. The ability of plants refers to growth and development.

3.4. TWO PROBLEMS

Here I must try to tackle two crucial problems with my reasoning. The first problem is an argument which says the following: When we are talking about the lowest level animals and the plants the only goals that we can find are the biological goals of survival of the individual or survival of the species. There are no other goals that could be termed vital among these living beings. Thus the biostatistical theory must come out as the only viable theory in this domain.

I first wish to dispute the general validity of this statement. It seems highly implausible that *all* strivings and *all* movements of low-level animals have the sole purposes of survival and reproduction. And if we admit that there may be at least some striving of such an animal that has another goal and that this striving can be frustrated by some inner process in the animal, then we have a case of illness that need not be explicable in biostatistical terms. Similarly, it seems highly implausible to say that the only valuable quality of a plant is its development for the purpose of surviving and reproducing itself. And if there are other valuable qualities and if they can be reduced by inner processes of the plant, then there is a case for talking about plant diseases in other terms than the biostatistical ones.

On the other hand, to cut the debate short, I can concede that to the most primitive living beings, such as the single-cellular ones, it is difficult to ascribe any further goals than the ones of being able to survive and to reproduce. Thus the diseases of such creatures would be well covered by the Boorsian concept. Hence, in such cases there is in practice a collapse between vital goals and certain biological goals. This, however, does not mean that the holistic theory fails to work in these cases. It is only that holism in these cases only contains certain biological goals. The welfare of a single-cell organism may be constituted only of its ability to survive and reproduce.

I think this point has to be emphasized. Survival and reproduction are not vital goals simply because of the biological fact that the organisms have a tendency to strive toward these states of affairs. Survival and reproduction are vital goals, as I see it, because they are inherent goods of the organisms in question. Thus, although there is a factual overlap between the biostatistical and the holistic viewpoint on this matter, there is a fundamentally different interpretation of the facts.

The second serious problem stems from a fundamental question. How can we determine the existence of a problem in a low-level animal or a plant? What is such reduced welfare of a low-level animal and a plant as is not identical with a lowered capacity for the survival of the individual or the species? Is there not a great risk of gratuitous ascriptions of goals and qualities to these creatures? Yes, there is such a risk. There is the particular risk of implanting *human* purposes in such animals and plants and indeed also in higher animals. Such health language already exists among veterinarians and agronomists. An animal or a plant may be deemed unhealthy because it does not fulfil the human expectations with regard to production, for instance of meat, milk or crop (West, 1995).

Indeed, when this health language is used, as for instance when we ascribe illness to a cow because it does not produce enough milk for us humans to consume, then we use a health-conception that is holistic. On the other hand this is not what I, and probably not what most of my holistic colleagues, *primarily* have in mind. Welfare, according to us, must be understood from the subject's point of view. A vital goal, in my theory, has to do with the subject's welfare. This is quite clearly indicated in the case of the higher animals, for which the crucial concept in my theory is the subject's long-term happiness. (A similar idea is proposed by Taylor (1986).)

In my preliminary discussion of animal and plant health above I was using the term 'vitality' in an attempt to cover the welfare of animals and plants. The Finnish philosopher Von Wright in a treatment of the concept of need in fact borrows the term 'flourishing' from the plant world to be applied to all living beings. A need of a living being, he says, is related to its ability to flourish, and not simply to its ability to survive. And, I would like to add, when such a need is frustrated because of processes internal to the animal or plant, then this living being is ill.

But how do we determine that vitality or flourishing is present from a subjective perspective when we have no real access to a subject's point of view? I do not deny the difficulty and profundity of this question and I have no illusion of coming up with an ultimately satisfactory answer. However, for the purpose of this paper there is an efficient short cut. I think my argument shows the following: To the extent that we are inclined to talk about the welfare of all animals and plants, we have a case for holistic notions of health and illness. Many people, including myself, are willing to ascribe vitality or flourishing to almost all animals and plants. We sometimes also observe how this vitality or flourishing can decrease. On this ground we can ascribe illness to the animal or plant. Our concept of illness is then a holistic one.

3.5. CONCLUDING REMARKS

Let me conclude this analysis. I have argued that, perhaps contrary to common belief, the so-called holistic theories of health are not confined to the human arena. We need not say that there is something special about human health that calls for a holistic theory of health, whereas the biostatistical theory or some equivalent theory must do the work in the case of animal health. I argue that health and illness could, or indeed should, be viewed in a holistic way with regard to all the living world. The concept of illness, I maintain, is related to the notion of a *problem* caused by internal processes. And all living beings may have such problems. These problems may be a threat to long-term happiness and this is typical for humans and the higher animals. However, in the case of non-sentient animals and plants the welfare goals must be different. In the limiting case they may be identical with survival and reproduction. Thus, they might in such extreme cases coincide with the biological goals of the biostatistical theory of health. This, however, does not reduce the holistic theory to the biostatistical theory in such cases. According to the holistic theory, as I see it, the goals of survival and reproduction constitute the welfare of the most primitive animals and plants.

Department of Health and Society, Linköping University, 58183 Linköping, Sweden

NOTES

[1] My preference concerns the basic concepts to be used. It will be evident later in my analysis that when we are to specify the nature of the abilities constituting health we will have to use notions such as well-being or welfare again. But then it concerns long-term well being and welfare and not the immediate well being that is at stake in feeling-oriented theories of health.

[2] Taylor suggests here that the well being of the organism is a means for preserving its existence. This may look as if his theory might in the end support a Boorsian analysis of health. However, what is crucial for my argument is that there is a notion of well-being of a lower level organism that is *not defined* in terms of mere survival. There is no indication that Taylor proposes such a definition.

REFERENCES

Barnard CJ, Hurst JL (1996) Welfare by design: The natural selection of welfare criteria. Anim Welf 5:405–433

Bennett ATD (1996) Do animals have cognitive maps? J Exp Biol 199:219–224

Boorse C (1997) A rebuttal on health. In: Humber J, Almeder R (eds) What is disease?. Humana Press, Totowa New Jersey, pp 1–134

Broom D (1998) Welfare, stress and the evolution of feelings. Adv Stud Behav 27:371–403

Bshary R, Wickler W, Fricke H (2002) Fish cognition: A primate's eye view. Anim Cognit 5:1–13

Canguilhem G (1978) On the normal and the pathological. D. Reidel Publishing Company, Dordrecht

Dawkins MS (1990) From an animal's point of view: Motivation, fitness and animal welfare. Behav Brain Sci 13:1–9

Duncan IJH (1996) Animal welfare defined in terms of feelings. Acta Agricult Scand Sect.A, Anim Sci Supplementum 27:29–35

Fulford KWM (1989) Moral theory and medical practice. Cambridge University Press, Cambridge

Galen (1997) Selected works, translated and introduced by P.N. Singer. Oxford University Press, Oxford

Griffin DR (1984) Animal thinking. Harvard University Press, Cambridge

Karavanich C, Atema J (1998) Individual recognition and memory in lobster dominance. Anim Behav 56:1553–1560

Lopez F, Serrano JM, Acosta FJ (1994) Parallels between the foraging strategies of ants and plants. Tree 9:150–152

Nordenfelt L (1995) On the nature of health, 2nd edition Kluwer Academic Publishers, Dordrecht

Nordenfelt L (2001) Health, science, and ordinary language. Rodopi Publishers, Amsterdam

Reznek L (1987) The nature of disease. Routledge & Kegan Paul, London

Schramme T (2000) Patienten und Personen: Zum Begriff der Psychischen Krankheit. Fischer Taschenbuch Verlag, Frankfurt am Main

Taylor PW (1986) Respect for nature: A theory of environmental ethics. Princeton University Press, Princeton

Temkin O (1963) The scientific approach to disease: Specific entity and individual sickness. In: Crombie AC (ed) Scientific change: Historical studies in the intellectual, social and technical conditions for scientific discovery and technical invention from antiquity to the present. Basic Books Inc., New York, pp 629–647

Toates F (1987) The relevance of models of motivation and learning to animal welfare. In: Wiepkema PR and van Adrichen PWM (eds) Biology of stress in farm animals: An integrative approach, pp 153–186

Wakefield JL (1992) The concept of mental disorder: on the boundary between biological facts and social values. American Psychologist 47:373–388

West GP (ed) (1995) Black's Veterinary dictionary, 18th edn. A & C Black, London

Von Wright GH (1963) The varieties of goodness. Routledge & Kegan Paul, London

Von Wright GH (1996) Om behov [On Needs]. In: Klockars K, Österman B (eds) Begrepp om hälsa [Concepts of health]. Liber, Stockholm

4. DISEASE AND THE CONCEPT OF SUPERVENIENCE

How to understand the social construction of disease

Within a wider debate about the status and nature of medical concepts, there is a viewpoint often called 'the social construction of disease.' I will focus on what this position precisely holds concerning the concept of disease and whether it supports, as it sometimes claims, being and anti-realism about disease and thus an anti-naturalist about all of medicine.

Realism and anti-realism are notoriously unclear positions. For purposes of this article I stipulate that a realist about the study of medicine (and specifically disease) believes there are medical facts and that it is in light of those facts that medical judgments can be said to be true or false. The reason this position is called realism is that the standing of the facts of the matter, namely what the world is like, is what makes the medical judgments true or false. These medical facts are or are not the case, whether or not anyone possesses the relevant concepts of illness or disease.

The term naturalism will also play a role in my discussion and it is, if anything, more vague and ubiquitous than realism. Again, I must simply stipulate it to be a worldview wherein only certain sorts of causal explanations of events are considered appropriate and a worldview broadly compatible with both science and much of common sense.

Social construction, in contrast, rejects the notion that medical theory and practices presuppose such a domain of medical facts about which true or false claims can be made. Some of these critics also reject the notion that medicine is part of the sciences, and some in addition reject the entire worldview of naturalism. For most constructivists, however, medical judgments are basically evaluative judgments and in that way are understood as expressing the values or commitments of institutional authorities and their social ideologies. The point is that medical judgments are not, in the final analysis, claims about the world at all, but only mask themselves as such as part of the exercise of some insidious ideological power. What is unmasked, the constructivist critic maintains, is that 'when we are deciding whether some condition is a disease, we are deciding what sort of people we ought to be' (Reznek, 1987, p. 166).

Naturalism and realism, however, are separate and distinct positions and therefore it is possible in such a debate to agree with the constructivist's claim that medicine involves normative claims and yet adhere to naturalism. But that would be to get ahead of matters at this point, and I want to wait until we have the social constructivist position on the table before raising questions about whether and how these two motivations of anti-realism and anti-naturalism coincide.

H. Kincaid and J. McKitrick (eds.), Establishing Medical Reality, 35–45.
© 2007 *Springer.*

I have two aims in this discussion. First, I want to present the position as generously as I can. Thus I plan to simply take on a good deal of the position. But to capture social construction in a way that seems to me to give it purchase in this debate, I find I need to introduce the concept of supervenience. Supervenience is a term of art in recent philosophy and therefore requires some further discussion below. I want to stress, however, that in so stating the position of construction I only intend to find a conceptual space for it. I am not claiming to refute or defend it merely by characterizing it through the concept of supervenience.

Therefore I want to proceed differently than those who, for example, engage in debates about whether disease is a natural kind (D'Amico, 1995, pp. 551–69). In those debates it is often supposed to fall out from a discussion of the concept of natural kind that diseases are fixed features of the natural world. This sort of conceptual analysis is supposed to then directly secure the place of medicine as a science. But the concept of a natural kind is itself contentious (Witmer and Sarnecki, 1990, pp. 245–64). Moreover, the concept of a natural kind has the effect of obscuring the social constructionist's approach from the start.

My second aim is to show, however, that the claim that diseases or illnesses are socially constructed either leaves this very debate about realism and naturalism in medicine wholly unresolved (contrary to what these critics intend) or the position turns out to be highly implausible.

This line of criticism, however, does not show that the doctrine of social construction suffers any fatal flaw. As a robust defense of the fundamentally evaluative nature of medicine, social construction can, at some cost I will argue, hold that medical knowledge is either illusory or more akin to claims in ethics and aesthetics than science. Though I believe there are less counter-intuitive ways to understand how the role of evaluative judgment in medicine co-exists with a broadly naturalistic account of medicine that would be too large a task for now. My aim is only to challenge the doctrine of social construction of disease, at least, as I understand it.

But the first problem is to choose a focus for the label of social construction. It is a phrase now used in a bewildering variety of ways and often mixing, for instance, anti-realist and skeptical conclusions (Hacking, 1999; D' Amico, 1998). Also some social constructivists in medicine see themselves as exclusively concerned with case studies and either ignore or just assume a more ambitious challenge to the entire discipline of medicine.

The term social construction seems to have been coined within sociology and psychology during the 1970s (Berger and Luckmann, 1966). It is now an umbrella for attacks on science and naturalism within the history of science in general and within the history and sociology of medicine specifically. Again, as noted above, the constructivist critics see science as an ideology and scientific research as a species of institutional power struggle.[1]

Instead of attempting to clarify the strategy first, I will begin with a typical example. My point is not to challenge the example per se. In fact, the case study chosen need not be at issue when addressing the larger question of the concept of disease. It does serve, however, to highlight some key features of the position.

Marie Rodin (1992, pp. 49–56) argues that the medical classification of premenstrual syndrome is a mistake due to the biases of a pervasive ideology within the medical community manifested in evaluative judgments about women supported by methodological defenses of medicine as a science. What Rodin finds behind this fake syndrome is then a family of prejudices throughout medical education and within medical institutions. As she states her aim; 'Conceptual and methodological problems inherent in contemporary biomedical research on PMS [premenstrual syndrome] are cited as examples of how medical knowledge is informed by Western beliefs and expectations about the relationship between the menstrual cycle and "irrational" and "uncontrollable" behavior in women' (Rodin, 1992, p. 49).

This article highlights two ways in which the strategy of social construction is used. First, it guides her case study because there is a syndrome falsely treated as a disease. In this way the article resembles many studies, some predating this label, where ideological, religious, or moral reasons are detected behind medical classifications. (Engelhardt, 1976, pp. 256–68; Szaz, 1961; Foucault, 1967.) In fact, the anti-psychiatry movement of the 1950s and 1960s seems an inspiration for these recent case studies.

But it is important to stress that, for instance, arguments against the reality of mental illness proceeded by contrasting it with paradigm cases of disease. Thus the anti-psychiatry critics were, at least initially, not denying the reality of disease.[2] I am not challenging these sorts of case studies (assuming that there is evidence of the mistakes) and I think that my overall conclusion challenging the more ambitious claims of social construction reinforces the potential contributions of such case studies.

The second strategy found in Rodin's article is the more ambitious claim. It holds that the very distinction between medical conditions classified as diseases and those conditions not so classified are illusory or merely an artifact of the ideologies she swishes to unmask and challenge. Rodin reaches this conclusion, I imagine, because she holds that all such classifications of medical illnesses and diseases are these evaluative judgments reflecting the prejudices or preferences of one's institution or community. It is this understanding of medical concepts that I am focusing upon in this paper.

Rodin is actually not clear which strategy her article supports and adopts. The reason I say it is not clear is because Rodin does not directly conclude, as one might expect at first, that PMS is not disease because of her evidence distinguishing its diagnosis and treatment from paradigm cases of disease. Rather, she shifts to the stronger claim that all such medical classification is ideological. For example, she implicitly makes the assumption that since all medical conditions require these sorts of evaluative judgments, then it is an ideological-political dispute as to whether or not PMS is a medical condition. It all reduces to the question of whose ideology shall predominate.

But I call this shift ambiguous because Rodin simultaneously marshals evidence of various matters of fact being ignored or misunderstood by the medical community (due, as she holds, to institutional evaluative biases). I will try to explain why this ambiguity, arising in so many articles similar to hers, is part of the position of social construction.

For now I will list, provisionally, the main features of claiming that diseases are socially constructed. I will accept these conclusions, except for the matters of emphasis and interpretation discussed below, because it will be my view that once properly understood they provide no definitive challenge to either a broadly naturalistic or realist account of medicine.

Constructivism holds that classifying conditions as diseases requires making evaluative judgments. It holds that evaluative judgments cannot be eliminated from the concept of disease. It holds that there are prejudices within the scientific conception of medicine dismissing or ignoring evaluative judgments within medicine. It holds that in disputes about whether a condition is or is not a disease, the disputes often involve value judgments. Finally, social construction introduces a causal story of how evaluative judgments are formed by, imposed by, and maintained by institutional social structures. This causal story involves discussion of belief-formation and such mechanisms as social-psychological conditioning.

How is an evaluative feature supposed to arise in the classifications or concepts of medicine? The concept of supervenience has been central for many decades in ethical theory, philosophy of science, and metaphysics. There are many different conceptions of supervenience and debates about whether and how these differing concepts are related, but my central point for now is much narrower and thus I hope to steer around these larger discussions.

The concept of supervenience was invented, as it might be put, to provide an alternative to the idea that evaluative properties reduce to non-evaluative properties (Hare, 1989, pp. 66–81). Hare has the clearest claim to having introduced the concept into philosophy in roughly the way that it is presently understood. But more important for my purposes, Hare explicitly intends that his characterization of what he calls 'supervenience phenomena' is neutral with respect to debates about realism and naturalism.[3] Thus Hare's approach allows me to state the position of social construction in such a way that the question of whether it then supports anti-realism or anti-naturalism can be dealt with separately.

Hare asks us to imagine two rooms of identical shape, furnishings, and orientations. He wants to then test our reaction to three statements. This first room is *nice*, but that second room (similar in all other respects) is not *nice*. This first room is *blue*, but that second room (similar in all other respects) is not *blue*. This first room is *hexahedral*, but that second room (similar in all other respects) is not *hexahedral*. With some further specification about 'other respects' that we can ignore, Hare argues that in one sense *nice* functions like *hexahedral* and in another sense it functions like *blue*.

First, it functions like hexahedral since both the statement using nice and the statement using hexahedral are self-contradictory, but not the statement using blue.

Second, it also functions like blue. Someone could give the full description of the room and conclude it is nice, and then change his mind and give the very same description and conclude it is not nice. The two statements are still contradictory, but they are not internally self-contradictory in the way that using hexahedral in its place would be. Hare's central point is that nice shares a logical restriction found in the hexahedral case, but is distinct from that case.

Hare thinks that we have that logical feature of restriction because 'supervenience requires that what supervenes is seen as an instance of some universal proposition linking it with what it supervenes upon' (Hare, 1989, p. 68). Thus, it does not permit (in the absence of a statement of one's change of mind) that we make different judgments of niceness about cases we admit are descriptively similar, because it is by some universal claim that reasoning can entail that the case in question has that property.

What is important here and thus connects with my previous discussion is that Hare does not require that the universal premise is a necessary truth; it might be a contingent, nomological claim. A universal premise is required to say that the feature in question supervenes. But then it may seem at first an oddity of Hare's view that the universal claims need not hold necessarily. Yet that point is part and parcel of why he introduces the term as he does. 'In morals ...some might say that universal moral principles hold necessarily; and others might prefer to say that they do not hold necessarily, but justify a necessary inference from themselves, plus a subsumptive premiss, to a moral judgement as conclusion. For myself, I prefer the second way of putting the matter. The important point is that supervenience by itself does not settle it one way or the other, either in the moral or in the causal case' (Hare, 1989, p. 76).

Hence the following case is possible. Someone could hold that medical judgments about health were the purported universal claims of authorities, applied with consistency, but in no sense necessary as a matter of the facts of the case. Supervenience implies, as presented above, a universal premise, but does not entail either logical or causal necessity for it. As Hare puts it, 'there is a sense in which things can be nice because they are as they are, without it being compulsory on us to like, and call nice, that kind of thing' (Hare, 1993, p. 75).

It would seem that this is just the idea of how the evaluative features within medical concepts are to be understood by constructivists. First, constructivists are led to think that statements about disease are non-reducibly normative rather than descriptive: there are disagreements in medicine about values that cannot be resolved by the matters of fact. The judgment that one has a disease is a judgment that does supervene on all of the properties of the patient, but this evaluative judgment about the patient is not compulsory and it could change without the facts changing.[4] Thus disputes about disease, like disputes about niceness, are not settled by agreement on what the facts of the case are.

Could showing that the concept of disease involves an evaluative judgment supervening on non-evaluative facts of the matter establish anti-realism about disease or that medicine is not a naturalistic discipline? I have already said that Hare's approach does not require that conclusion and I don't think the success in this criticism of medicine can be so easily attained. In fact I want to show that there is a version of social construction that provides no obstacle at all to taking diseases as real and medicine as a naturalistic discipline. Thus whether correct or not in any of its specific analyses, it would be as a position idle with regard to these larger debates about realism and naturalism.

I want to then show that the cost of making the position one that does require rejecting realism about disease or naturalism in medicine produces a highly implausible version of social construction. It makes the view much less plausible than the version idle with respect to the debate, and much less plausible than either a realist or naturalistic account.

I will begin with the version I claim is plausible but idle with respect to the debate. What is important in this context is to focus on the causal analysis of the formation of beliefs and attitudes that gives the position the name social construction. It should be understood as showing that whatever is picked out by the concept of disease, these features, physical states, sensations, feelings, preferences, desires, and beliefs, are causally interacting.

Though it is a large task to defend and of course beyond this paper, there is at least a plausible argument that features covered by causal explanations such as these are candidates for features of the natural world. Also it is worth stressing that the constructionist's emphasis on conditioning and indoctrination need not exhaust the ways in which mental states causally interact with the rest of the natural world. The causal analysis of the formation of beliefs and attitudes would have to be open to whatever are the contingent facts about how psychology operates.

Furthermore, nothing about this constructivist causal analysis of mental sensations, attitudes and beliefs requires a commitment to scientific reduction with respect to beliefs, desires, and preferences. This version of construction is compatible with there being real mental states of these sorts not reducible to physical states. Such an emergent realism with respect to all of these psychological matters is compatible with much of social constructionism and yet also compatible with naturalism, though perhaps not with all forms of scientific realism.

A naturalist will think that since states of being diseased are playing these causal roles within the world that there are some natural properties that the concept of disease picks out. Now having said that I do not of course think that proves there are such natural properties. I am merely holding that given the totality of beliefs about the functioning of the natural world as they stand today, it is plausible to think that the causal story within social construction is supporting evidence for naturalism in medicine and perhaps also realism about disease.

But I did stress that the position of realism is separate from naturalism. How might that point be addressed? Since medicine, if it is a science, is a social, biological science, it is best not to craft physical science analogies for or against the case of medical realism. Therefore I will use pain phenomena as my example.

The study of pain requires the study of sentient beings. Therefore, in a strict sense, there are no pain phenomena in the same way that there are heat phenomena in the physics. But it would be confusion based on this analogy to then conclude that therefore pain is not real or that our concept of pain is massively in error.

Understanding the natural causal order for the phenomena of heat in physics – and of course its relation to our common sense conception of heat – does not require including sentient beings in that causal network. Therefore, even though we concede that without sentient beings there would not be pain that concession simply treats the causal generation of pain as more complex. In effect, being in pain supervenes upon the totality of facts about a certain sort of being.

For these reasons, the causal analysis characteristic of case studies supporting social does not require agreeing with its anti-realist and/or anti-naturalist conclusions. The position is simply neutral with respect to those conclusions, at least as so stated.

Could the constructivist modify the position in such a way that it does support anti-realist or anti-naturalist conclusions? The only way I can see is for the constructivist proponent to make the case that there are no disease properties in the world. If such properties are not the case, then their study could not of course be naturalistic. These critics could agree that the language of medicine expresses favorable or unfavorable attitudes toward the world and may even prescribe actions. None could turn out to be true or false nor would medicine be said to identify features of the world that either are or are not the case. The idea of medical research and investigation is therefore an error or a sham.

The first point that can be made is that such an extreme view is still compatible with naturalism. As I pointed out earlier, it looks like some constructivists are pointing to how the concept of disease and perhaps all concepts of medicine differ from the rest of science. Thus our being massively wrong about the nature of medicine as a discipline might well be motivated by the intent to preserve the naturalistic conception of science.

But how plausible is the idea of our being massively in error about the nature of medicine? I think it is hugely implausible. Of course there are examples of other disciplines about which it could be said that we might be in massive error about the nature of that discipline. Perhaps aesthetic judgment about objects being ugly or beautiful is a case in point. Perhaps we are picking out no such thing as the property of being beautiful because there really is no such property. We would then need to explain what we are really doing when we claim to be picking out those features of objects and perhaps some larger story about human psychology could explain both our persistence in making aesthetic judgments and our propensity to treat them in this way. Perhaps the same could be said of theology.

But the aesthetic example works only because there are independent reasons to think there is no property of 'beautifulness.' We think so because of the poverty of evidence for the right sort of causal regularities of such aesthetic matters (at least those not reducible to regularities of psychology alone). Such additional features of the world are not needed; at least as things now stand.

But when we turn to the example of disease, for instance, there is considerable evidence of regularities holding even as medical concepts and values change.

Specifically, there are regularities that are species-specific and significantly there are causal interactions between being in a disease-state and entities that are uncontroversially natural and real (for example, viruses, bacteria, and solar or nuclear radiation). Also, this persistent evidence of regularities cannot, without exaggerated gerrymandering, be subsumed under psychological regularities alone. Unlike the case of the aesthetic properties, which of course might for all that poverty of evidence nevertheless turn out to be real, explanation of how there could be such massive error about medicine looks far-fetched.

If the study of disease is part of the biological sciences, then it is the study of beings of a certain sort about whom one can also make, for instance, evaluative judgments concerning their purpose or good. But that is not a reason to think medicine is non-naturalistic even while biology is naturalistic. But the constructivist might still insist that medical judgments do diverge from the rest of biology. No matter how detailed knowledge of liquidity or cellular metastasis might become, the action-guiding feature of the concept of disease (which is the feature of an evaluative judgment that such and such is a bad condition for such a being) must be presupposed by any version of naturalism.

For example, we might ask how medicine could make the distinction, on the facts alone, between bad conditions that are diseases and bad conditions that are not (Hare, 1993, pp. 31–49). How do we distinguish between male-pattern baldness (not a disease or even a condition requiring medical attention) from alopecia areata (a disease whose symptoms involve hair loss occurring in a certain pattern of clumps)? Alopecia areata has been hypothesized to be a form of autoimmune disease of the hair follicles. Is there a way to hold that we do have evidential grounds for this distinction?

First, there may be some notion of species-design entering into the medical distinction between these cases since the fact that the alopecia afflicts both males and females turns out to be a relevant consideration about whether it is a disease (Boorse, 1975, pp. 49–68; 1977, pp. 542–73). Second, the features of the hair loss distinguish the two cases. Finally, though it is an evaluative judgment to say that hair loss is bad, there is a difference between a medical and aesthetic judgment of what is bad.

Hare objects to such a defense, for instance, because he objects to treating baldness as a failure of species-design.[5] But nothing he says about that example disallows treating the badness of alopecia as distinct from the badness of inherited baldness (or as distinct from the act of shaving one's head). Perhaps some think male-patterned baldness is bad for them, but perhaps they do so for aesthetic reasons. Hence there is an evaluative judgment not proper to a medical context, and not grounded by medical justification. Similarly one might consider the length of his nose as bad for him, where bad is again being used non-medically. He could of course seek surgery. But the large nose in this case is not a medical condition and its treatment not a response to a disease or illness.

At this point the committed constructivist will simply insist that the distinction between these cases breaks down, in the end, because it is not based on any matter of fact. Even were there discovered a causal agent of alopecia (a virus, for example), its being classified as a disease (in distinction from male-pattern baldness) remains wholly independent of that fact, were it a fact, and its being so classified is simply presupposed by medical theories and institutional practices.

But let me return briefly to the alopecia example. I have granted that in saying alopecia is a disease there is some reliance on an evaluative claim of the sort that such a condition is not a good medical condition for such a being. Why don't we extend that judgment all baldness? It would seem that what motivates not extending the judgment is simply the sum-total of evidence. For example, alopecia's symptoms suggest a discrete causal agent (in contrast with genetic inheritance), it afflicts males and females, and its symptoms are distinct from male-patterned baldness.

But what about other inherited diseases? Would not they then be similar to male-patterned baldness? These are of course complex questions. In some cases one inherits susceptibility to various external agents. Also, contrary to our current non-lethal example, normally inherited diseases are restricted to lethal conditions for those sorts of beings. Inheriting perfect musical pitch is simply not a medical matter at all, so inheritance alone is never enough evidence.

As in any naturalistic scientific inquiry there will be a process of adjustment and critical examination of the common sense and scientific vocabularies over time. Therefore I am engaging in speculation here. I am merely sketching what I mean by saying it is plausible that the social constructivist's story does not exclude the realistic attitude toward medical concepts and how the accumulation of evidence might reinforce such a conclusion. There may well be many points where our common sense picture clashes with medical classifications or where it obscures classifications, but such matters present no insuperable obstacle for either naturalism or realism about diseases.

What if some constructivists are moral relativists? Perhaps the point is that the values arising in their causal stories are relative to communities or societies. Then what we ought to do about our bodies or what kind of people we ought to be will diverge and change over time wholly independent from whatever are the matters of fact.

Such relativism, whether acceptable or not, need be neither anti-realist nor anti-naturalistic. After all, it could well be that some of the disputes are just as these constructivist say they are. For example, Engelhardt's (1976, pp. 256–68) discussion of the history of medical disputes about the classification and treatment of masturbation concludes that the entire topic was about cultural values and no fact at all. But his conclusion emerges from the details of the case study and leaves it open, of course, that there are different cases. As Rodin suggests, some cases are ones where the key matters of fact were not available or were ignored or misunderstood.

In conclusion, the recognition that there are evaluative claims in medicine does not preclude such values turning out to be constant among certain beings precisely in virtue of their underlying biology. But, if so, then medicine has at least a case to make that it investigates those matters of fact. It investigates the regularities with respect to what conditions sustain and erode our underlying biological-psychological well being.

ROBERT D'AMICO

Department of Philosophy, University of Florida, Gainesville, Florida, USA

NOTES

[1] Michel Foucault (1973) speaks of disease as merely a name picking out no real thing or essence and describes his work as a 'nominalist reduction' of the concept. Francois Delaporte (1986) begins by asserting that diseases do not exist; only social practices about diseases exist. Lawrie Reznek (1987) is unique in defending medical anti-realism in detail. Both Foucault and Delaporte envisage medical studies shifting to a sociology and psychology of institutional, collective behavior.

[2] Foucault (1967) began his career arguing that mental illness was not a disease and therefore that the medical treatment of the insane was morally wrong and perhaps politically motivated. Mental illness, in his controversial history, is merely a 'tool' for the nation-state's effort to control dispersed populations of poor vagrants. But setting aside whether he has evidence for this political conception of madness, Foucault lacks reasons for distinguishing being diseased from being mentally ill. He only offers historical evidence of the errors, prejudices, and indignities produced by past (and present) theories and treatments of mental illness. But such historical evidence could not constitute evidence that madness is not a disease, since similar historical errors and confusions are found with regard to non-mental disorders. Such historical facts are consistent, in other words, with the disorder turning out to be a disease. In *The Birth of the Clinic* (1973) Foucault resumes the debate so as to block the above objection. Foucault declares the entire concept of disease a conventional classification. One can see the rationale since, as I pointed out, there is evidence of conceptual confusion and improper treatment in cases of physical illness. But historical evidence alone cannot establish, as his earlier restricted argument about madness also failed to establish, that the very idea of a disease is illusory.

[3] 'My own first use of the word …was in the course of an attempt to find clear logical criteria for distinguishing between evaluative and purely descriptive words …in order to refute ethical descriptivism … But I came to think that [was] …mistaken. For supervenience is a feature, not just of evaluative words, properties, or judgments, but of the wider class of judgements which have to have, at least in some minimal sense, reasons or grounds or explanations. And, as we shall see later, it is possible, at a certain cost in queerness, to remain a descriptivist or a realist, even of a non-naturalist stamp, and still believe in supervenience.' (Hare, 1989, p. 66) In Hare's prose, I will pay a 'cost in queerness' defending medical realism, if the concept of disease is evaluative. But I am trying to separate this issue from whether moral realism or anti-realism is correct. Hare concedes that both the realists and anti-realists are free to adopt the notion of the evaluative supervening on the descriptive, and that will be enough I think to show why realism about the concept of disease is plausible. But I am not thereby making moral realism plausible as well.

[4] Perhaps this clarifies what Foucault means when he enigmatically states that the 'superposition of the …disease and the body of the sick man is no more than a historical, temporary datum.' (Foucault, 1973, pp. 3–4) Superposition seems a stand-in for supervenience in that, for Foucault, instantiation of all the physical facts about the patient does not necessarily fix the patient being diseased, or only does so within that historical context.

[5] Hare challenges Boorse's attempt (Boorse, 1977) to define disease in terms of 'natural function' and its species-specific contributions to survival or reproduction. But Hare objects; 'The growing of hair …seems to be a natural function, and it seems that a condition which prevented it might …be called a disease …but baldness …is not inimical to survival or reproduction. The reason why we call conditions causing it diseases is simply that people do not *like* being bald' (Hare, 1993, 40–41).

REFERENCES

Berger PL, Luckmann T (1966) The social construction of reality: A treatise on sociology of knowledge. Garden City, Doubleday

Boorse C (1975) On the distinction between disease and illness. Philosophy and Public Affairs 5:49–68

Boorse C (1977) Health as a theoretical concept. Philosophy of Science 44:542–573

D'Amico R (1995) Is disease a natural kind? J Med Philos 20:551–69

D'Amico R (1998) Spreading disease: A controversy concerning the metaphysics of disease 20:143–162

Delaporte F (1986) Civilization and disease. MIT Press, Cambridge

Engelhardt T (1976) Ideology and etiology. J Med Philos I:256–268

Foucault M (1967) Madness and civilization. Tavistock, London

Foucault M (1973) The birth of the clinic. Pantheon Books, New York

Hacking I (1999) The social construction of what? Harvard University Press, Cambridge

Hare RM (1989) Essay in ethical theory. Supervenience Oxford, Clarendon Press, pp 66–81

Hare RM (1993) Essays on bioethics. Health Oxford, Clarendon Press, pp 31–49

Reznek L (1987) The nature of disease. Routledge, New York

Rodin M (1992) The social construction of premenstrual syndrome. Social Science and Medicine 35:49–56

Szaz TS (1961) The myth of mental illness: Foundations of a theory of personal conduct. Dell Press, New York

Witmer G, Sarnecki J (1990) Is natural kindness a natural kind? Philosophical Studies 90:245–264

5. DECISION AND DISCOVERY IN DEFINING 'DISEASE'

Two weeks ago, Mr. Smith caught a cold. His nose cleared up in a week, but the cough continued and a few days ago it got worse. Last night he couldn't sleep because he was coughing so much, and for the first time he developed a fever and chills. Today his temperature is 101 F, he is coughing up some brownish sputum, and he feels short of breath. He has pneumonia – confirmed by the physical exam and chest x-ray – and is admitted to the hospital for antibiotics, fluids, and observation.

5.1. INTRODUCTION

The debate over how to analyze the concept of disease has often centered on the question of whether to include a reference to values, in particular the 'disvalue' of diseases, or whether to avoid such notions. Using Mr. Smith's condition as an example, it is easy to see the attractive aspects of each approach. 'Normativists,' such as King (1954) and Culver and Gert (1982) emphasize the undesirability of diseases and the harms and limitations they bring, and Mr. Smith's condition certainly has these features. 'Naturalists,' most prominently Christopher Boorse (1977, 1987, 1997), instead require just the presence of biological dysfunction, and again Mr. Smith's case fits the theory: the infection is now interfering with his lungs' ability to carry out their function of absorbing oxygen and releasing carbon dioxide.

The debate between normativism and naturalism often deteriorates into stalemate, with each side able to point out significant problems with the other. It starts to look as if neither approach can work. In this paper, I argue that the standoff stems from deeply questionable assumptions that have been used to formulate the opposing positions and guide the debate. In the end, I propose an alternative set of guidelines that offer a more constructive way to devise and compare theories.

In Section 5.2, I describe some of the current methodological precepts and provide a way of classifying theories that avoids the classic dichotomy of normativism vs. naturalism. In Sections 5.3 and 5.4, I present an overview of the strengths and weaknesses of four of the most important analyses of the concept of disease. In Section 5.5, I argue against the idea that there is just one correct definition to be discovered and that this definition must be a short list of necessary and sufficient criteria. These assumptions – part of the project of *conceptual analysis* according to many philosophers – are undermined by psychological research on concept-use and philosophical arguments about meaning.

47

H. Kincaid and J. McKitrick (eds.), Establishing Medical Reality, 47–63.
© 2007 *Springer.*

Once these assumptions have been dropped, the way is open for a more productive approach, which I outline in Sections 5.5 and 5.6. In particular, the new framework allows us to give adequate importance to both values and biological dysfunction in defining 'disease.'

5.2. CLASSIFICATION AND METHODOLOGY

5.2.1. Classifying Theories

The first step in disentangling the debate is to move beyond the terminology of 'naturalism' vs. 'normativism.' Instead, I will use the term 'value-requiring' (VR) for theories that define 'disease' using concepts such as 'disability,' 'harm' or 'evil.' Theories relying on such ideas include Lester King's (1954) account, which defines diseases as conditions that are judged by society to be 'painful or disabling' (p. 112). Culver and Gert (1984) similarly invoke values but drop King's relativist approach, focusing instead on the objective features that make diseases undesirable to many.

For the opposing side, instead of 'naturalist,' I prefer the more precise 'non-value-requiring' (non-VR). Christopher Boorse's (1977, 1987, 1997) theory, for instance, is non-VR, since he includes no criterion involving the value or disvalue of a condition. Instead, his theory requires just the presence of biological dysfunction, and thus I will call the account 'dysfunction-requiring' (DR). King (1954) and Culver and Gert (1984), in contrast, do not refer to the presence of dysfunction, so their accounts are 'non-dysfunction-requiring' (non-DR). The benefit of this more precise terminology becomes apparent when we consider Jerome Wakefield's theory, which requires both the presence of dysfunction and the tendency to cause harm. His theory is thus both dysfunction requiring and value requiring (DR and VR). The advent of a theory such as Wakefield's helps illustrate how the normativist/naturalist divide is starting to break down (see also Richman, 2004).

I'll focus on these four theories: King (1954), Culver and Gert (1982), Boorse (1977, 1987, 1997), and Wakefield (1992a, b, 1999a, b). Their definitions are listed in Table 1. The theories can be classified by using a 2×2 grid (Table 2), with the horizontal axis divided between dysfunction-requiring (DR) and non-dysfunction-requiring (non-DR), and the vertical axis divided between value-requiring (VR) and non-value-requiring (non-VR). The theories thus fall in three of the four squares of this grid, and I will take the strengths and weaknesses of each as representative of these general approaches.

5.2.2. 'Disease' and Methodology

All these accounts offer their definitions as necessary and sufficient criteria for a condition's being a disease. In so doing, they reflect what has been called the 'classical view' of concepts, where definitions are stated by a short list of individually necessary and jointly sufficient criteria for their application (Ramsey, 1992). As mentioned above, the classical view is widely assumed in analytic philosophy, as

Table 1. Definitions of Disease

King (1954): Disease is the aggregate of those conditions which, judged by the prevailing culture, are deemed painful, or disabling, and which, at the same time, deviate from either the statistical norm or from some idealized status. (King, 1954, p. 197)

Culver and Gert (1982): A person has a malady if and only if he has a condition, other than his rational beliefs and desires, such that he is suffering, or at increased risk of suffering, an evil (death, pain, disability, loss of freedom or opportunity, or loss of pleasure) in the absence of a distinct sustaining cause. (p. 81)

Boorse (1997):

1. The *reference class* is a natural class of organisms of uniform functional design; specifically, an age group of a sex of a species.
2. A *normal function* of a part or process within members of the reference class is a statistically typical contribution by it to their individual survival and reproduction.
3. A *disease* is a type of internal state which is either an impairment of normal functional ability, i.e. a reduction of one or more functional abilities below typical efficiency, or a limitation on functional ability caused by environmental agents.
4. *Health* is the absence of disease. (pp. 7–8)

Wakefield (1992a, p. 384): 'A condition is a disorder if and only if (a) the condition causes some harm or deprivation of benefit to the person as judged by the standards of the person's culture (the value criterion), and (b) the condition results from the inability of some internal mechanism to perform its natural function, wherein a natural function is an effect that is part of the evolutionary explanation of the existence and structure of the mechanism (the explanatory criterion).'

---

can be seen in debates over the proper conceptual analysis of concepts ranging from 'cause' in metaphysics, to 'justification' in epistemology, to 'justice' in moral theory.

The four theories to be considered here vary in their choice of terms – King and Boorse use 'disease,' Culver and Gert use 'malady,' and Wakefield uses 'disorder' – but all treat the relevant notion as encompassing all pathological conditions. As Boorse writes, in many contexts the term 'disease' applies to only a subset of conditions he is interested in, leaving out such things as injuries and malformations. But he says that he means 'disease' in a broad sense, taken as analogous to 'a pathologist's concept of disease' (1997, p. 11). Culver and Gert, likewise, say that they are not interested in the 'arbitrary' distinctions between concepts such as 'disease,' 'illness,' 'injury,' 'defect,' and 'injury,' aiming instead to encompass them all with their term 'malady' (1982, p. 65). And while Wakefield is mostly concerned with the concept of 'mental disorder,' he writes that he is interested

Table 2. Classification of Theories

	Dysfunction-requiring (DR)	Non-dysfunction-requiring (non-DR)
Value-requiring	Wakefield (1992a)	King (1954) Culver and Gert (1982)
Non-value-requiring	Boorse (1997)	

in a notion that applies to physical disorders as well (1999a, p. 376). I'll use the term 'disease' for the concept in question, while deferring to the terms used by the individual theories in appropriate places.

Each definition is generally supported by arguments that it does a better job than the others at classifying conditions as diseased or healthy. These arguments generally claim that an alternative approach incorrectly classifies some specific healthy condition (or conditions) as diseased, or some disease (or diseases) as healthy. Unfortunately, some of the key debates in the naturalist vs. normativist debate have involved imaginary conditions and unsupported intuitions about what doctors or others would or should say about them. Many projects of conceptual analysis in analytic philosophy rely on such imaginary cases and armchair intuitions – think of the convoluted and occasionally physically-impossible cases used in the debate over causation or personal identity – and while such cases can be useful, I believe they must be used with care. I will focus on arguments where historical or contemporary examples are used and where claims about classification are supported by good arguments or actual practice.

5.3. VALUES AND DISEASE

5.3.1. Culturally-Relative Values

Lester King presented one prominent value-requiring (VR) definition of 'disease' 50 years ago, writing,

Disease is the aggregate of those conditions which, judged by the prevailing culture, are deemed painful, or disabling, and which, at the same time, deviate from either the statistical norm or from some idealized status. (1954, p. 197).

The relevant values here are culturally relative, based on what the 'prevailing culture' believes, and there's much to recommend this sensitivity to cultural judgments. Historians and sociologists of medicine often focus on moral beliefs and value judgments in analyzing how societies decide which conditions count as normal and abnormal. Such analyses often uncover a key role for value judgments and other beliefs (cf. Rosenberg, 1992, pp. xii–xxvi).

For VR theories, though, adopting a relativist notion of value also carries negative consequences. In particular, it makes it difficult to explain how societies can be *mistaken* about what is a disease. Masturbation was considered a disease in 19th century America (Engelhardt, 1974, pp. 234–248), as was the desire of slaves to escape, labeled 'drapetomania.' It's not just that we have now stopped counting masturbation or a desire for freedom as diseases, but that we believe that *ever* classifying them as diseases was deeply mistaken. Similarly, some conditions previously considered normal and even desirable are now judged to be pathological. Take, for instance, the traditional Chinese practice of foot binding, which often left its subjects unable to walk. Classifying masturbation as a disease or deformed feet as normal are mistakes, like believing that the world is flat or that spirits cause hurricanes.

Relativists can respond in two ways. They might articulate and defend a wide-ranging relativism, where even the shape of the earth is a relative matter. More commonly, they will distinguish between such 'objective' facts as the shape of the earth and the value-laden judgments concerning states of the body. The most nuanced commentators pursue a sort of middle way. Charles Rosenberg, for instance, rejects the idea that societies 'construct' disease, instead arguing that they somehow 'frame' it and 'negotiate' its bounds (cf. Rosenberg, 1992). I won't consider the realism/ relativism debate in this area any further here, having at least pointed out the challenges facing a truly relativist approach.

5.3.2. Objective Values

Other VR accounts avoid relativism by defining the relevant value-laden notions in a non-relativist way. Culver and Gert's theory is representative of this approach. They choose the term 'malady' for their target notion and define it as follows:

> A person has a malady if and only if he has a condition, other than his rational beliefs and desires, such that he is suffering, or at increased risk of suffering, an evil (death, pain, disability, loss of freedom or opportunity, or loss of pleasure) in the absence of a distinct sustaining cause. (Culver and Gert, 1982, p. 81)

Here the relevant value-concept, 'suffering ... an evil,' is precisely defined in terms of a list of possible events – 'death, pain, disability, loss of freedom or opportunity, or loss of pleasure.' According to their theory, if masturbation or desiring to escape slavery does not actually increase a person's risk of experiencing death, pain, etc., then these conditions are not maladies, no matter what society believes. This avoids the most serious challenge facing the relativist version of VR theories.

That said, Culver and Gert's definition still allows too many conditions to count as diseases. In a society with certain beliefs, it may very well be the case that a slave's having a deep desire to escape does carry significant risk of death, pain, etc. In a society with certain beliefs, again, masturbating or rejecting foot binding will similarly carry real costs in terms of freedom, opportunity, or pleasure. Thus, Culver and Gert really just lengthen the path from societal beliefs to disease-status by one step: the value judgments now determine what counts as disease by way of their influence on people's lives. Similarly, it seems that in a bigoted society having dark skin or appearing to be a member of a certain ethnic group might count as a malady.

Critics have also pointed out that Culver and Gert's theory counts pregnancy as a malady because of the associated disability and the risk of complications and even death. Culver and Gert explicitly accept this conclusion (Gert, Culver, Danner-Clouser, 1997, p. 126), and this suggests that the concept of malady may differ significantly from the concept of disease that is actually at work in biology or medicine. From a biological point of view, of course, pregnancy and the process of reproduction is a normal state of the organism, no matter how risky. And although medical professionals treat women who are pregnant and in labor, it's not the pregnancy itself, but rather the conditions that may arise, which count as pathological. Health professionals providing contraception and carrying out abortions generally do not believe that pregnancy is a pathological condition.

While these problems for Culver and Gert stem from harm's being a sufficient criterion for disease, harm's being a necessary criterion as well causes other problems. For example, a disease may actually benefit an individual who has it, as when flat feet keep a person out of the army, or when an infection with cowpox confers resistance during a smallpox epidemic (Boorse, 1977, pp. 544–545). More fanciful examples are possible, such as a case where a researcher in a totalitarian country saves a person's life to study his mild eczema (following Boorse, 1997, p. 88). In these situations the conditions are still diseases, it seems, just ones that happen to be beneficial.

A VR account could respond in a couple of ways, for example by emphasizing that all these conditions carry some harm that is being counterbalanced by certain benefits. But allowing a condition to count as a disease just as long as it carries *some harm* – independent of the net outcome – would water the condition down to triviality, since all conditions carry at least some harm or risk of harm. Winning the lottery, for instance, carries risks inherent in acquiring so much money at once, wondering who your real friends are, etc. Another approach available for VR theories would be to define some environments as normal and others as abnormal, requiring that the condition causes net harm only in normal environments. This way forward, however, depends on a concept of 'normal environment' that appears as difficult to define as the notion of disease.

5.4. DYSFUNCTION AND DISEASE

5.4.1. Boorse's Account

Partially in response to the problems involved with making values necessary or sufficient to a condition's being a disease, Boorse (1977, 1987, 1997) proposes entirely eliminating value judgments from the definition. Instead, he focuses exclusively on whether there is biological dysfunction present. He states his 'Biostatistical Theory' (BST) as follows:

1. The *reference class* is a natural class of organisms of uniform functional design; specifically, an age group of a sex of a species.
2. A *normal function* of a part or process within members of the reference class is a statistically typical contribution by it to their individual survival and reproduction.
3. A *disease* is a type of internal state, which is either an impairment of normal functional ability, i.e. a reduction of one or more functional abilities below typical efficiency, or a limitation on functional ability caused by environmental agents.
4. *Health* is the absence of disease. (1997, pp. 7–8)

Each of the first three premises is central to applying his dysfunction-requirement. First, organisms are compared just with other individuals of similar age and same gender, i.e. members of the same 'reference class' defined in Premise 1. This is because levels of functioning that count as normal vary over age and between the genders in crucial ways. Human children younger than six months old can't walk and newly hatched chicks can't fly, but their legs and wings, respectively, do not count as dysfunctional. In mammals, the female breast has the function of producing milk after childbirth, while the male breast never does. Note that even value-requiring accounts may need to rely on some form of 'reference classes,' for example to explain when an individual has an 'increased risk' of suffering an evil (Culver and Gert, 1982, quoted above).

Second, a trait's function is defined (in Premise 2) as its 'statistically typical contribution' to survival and reproduction in individuals with that trait. Thus the eyes have the function of allowing sight, the heart has the function of pumping blood, and the lungs have the function of absorbing oxygen and releasing carbon dioxide. Philosophers differ about the best definition of 'function,' with some linking the notion to the evolutionary history of the relevant trait (cf. Neander, 1991, pp. 168–184). I won't address the debate between these accounts and Boorse's here, beyond noting that both work equally well in classifying Mr. Smith's case of pneumonia as disease, since both assign the lungs the same function. The lungs contribute to survival and reproduction by absorbing oxygen and releasing carbon dioxide, and they clearly arose under selection at least partly for carrying out these roles.

Third, the idea of a condition's being an 'impairment of normal functional ability,' in Premise 3, is meant to be a roughly statistical distinction. If a trait's level of functioning falls in the lowest percentile for people of the same gender and age, then it counts as dysfunction. In this way, the definition is meant to avoid the notions of disability or harm that value-requiring accounts rely upon. Mr. Smith's pneumonia counts as a disease, according to this account, since his lung function has decreased significantly. His lungs are carrying out enough gas exchange to keep him alive but are functioning at levels that are statistically unusual for men of his age. And it is this diminished functioning that makes Mr. Smith's case of pneumonia a disease, not the suffering or risks that it imposes.

5.4.2. Challenges to Boorse

Boorse's theory has been the most prominent non-value-requiring account and thus has attracted a great deal of criticism, which Boorse has recently addressed in a monograph-length paper (1997). Perhaps the most serious attacks on the account focus on the claim that dysfunction is a *sufficient condition* for disease. For example, if there is some part of the human brain that produces sexual attraction for members of the opposite sex, then this part would have the function of so doing, since this role is its statistically typical contribution to reproduction (as in Premise 2). But then it follows that if this part of a person's brain fails to produce such attraction in a small percentage of the population, resulting in homosexuality, the condition would count as a case of disease, by Premise 3. And this clearly contradicts the opinion of the medical profession and most people in the United States.

Boorse has bitten the bullet on this, accepting that it may turn out that his account makes homosexuality a disease, and he defends the theory by pointing out that the fact that a condition is a disease carries no inevitable implications about whether it should be treated by doctors (1997, pp. 11–13). Infertility would also be a disease, according to his theory, but a person who preferred to remain childless might understandably reject treatment as well. As mentioned in Section 5.2, Boorse says that he's interested in defining the theoretical notion of disease at work in medical science, leaving decisions about whether to intervene to people's free will. And these decisions, he accepts, will involve the application of values. He proposes that there might even be a different concept, which he calls 'therapeutic abnormality,' that should be defined by adding a value-requirement to his definition of 'disease' (1997, pp. 12–13).

Other challenges to Boorse's account object to the claim that dysfunction is a *necessary* criterion for a condition's being a disease. For example, it may be that female orgasm makes no specific contribution to survival or reproduction, and thus the mechanisms that bring it about have no biological function. But at the same time, a woman's inability to orgasm may be a serious problem for her, and one which physicians should treat as a disease (Reznek, 1987, p. 131). Again, Boorse responds by appealing to the distinction between a condition's disease-status and questions about whether medical professionals should diagnose and treat it if possible. Even if anorgasmia is not a disease, he writes, the suffering it causes may be a sufficient reason for health professionals to attempt to alleviate it. While the treatment of disease makes up a central part of medical practice, he writes, there may well be other conditions, which are not diseases, that health professionals have the skill and training to treat (Boorse, 1997, pp. 90–99). Doctors have long been involved in inducing sterility and fixing ugly noses, but they do so without claiming that fertility or ugliness are diseases. The cost of this approach, though, is to separate Boorse's definition of 'disease' from any clear conclusions about what medical professionals should do, and these risks making his definition look somewhat irrelevant to medical practice.

Both these problems for Boorse's theory – classifying homosexuality as a disease and anorgasmia as healthy – stem from the account's focus on the way that a trait affects just survival and reproduction. So even though homosexuality may be consistent with, or conducive of, pleasurable, loving lives, the condition counts as a disease because of its interference with certain ways of reproducing. And even though sexual satisfaction may be essential to many people's happiness, a condition like anorgasmia only counts as a disease if it interferes with reproduction. The focus on reproduction looks like it leaves out too much, i.e., all the other crucial goals and goods of human life. Of course, those are exactly the goods that VR accounts like King's and Culver and Gert's attempt to include, and we saw the problems with including so much. It starts to look as if the attempt to define disease faces a sort of dilemma: either include too much or too little of normal human goals.

5.4.3. Wakefield's Account

Jerome Wakefield (1992a, b; 1999a, b) attempts to answer some of these problems by combining a dysfunction-requirement with a value-requirement. He states his 'Harmful Dysfunction' account as follows:

A condition is a disorder if and only if (a) the condition causes some harm or deprivation of benefit to the person as judged by the standards of the person's culture (the value criterion), and (b) the condition results from the inability of some internal mechanism to perform its natural function, wherein a natural function is an effect that is part of the evolutionary explanation of the existence and structure of the mechanism (the explanatory criterion). (1992a, p. 384)

The theory thus avoids Boorse's apparent misclassification of homosexuality: even if homosexuality results from a biological dysfunction of some part of the brain, it is not a disease if it doesn't impose harm. Since the presence of dysfunction is necessary but not sufficient in Wakefield's account, it can avoid some of the problems facing Boorse.

That said, Wakefield's theory still has difficulties that stem from dysfunction's being a necessary condition: female anorgasmia will still not count as disease if orgasm has no function in women. And there are cases that do not involve sexuality or reproduction that can be used to attack the dysfunction-requirement in similar ways. (These examples challenge Boorse's account just as much, of course, since he treats dysfunction as necessary as well as sufficient.) Consider Mr. Smith's case of pneumonia again, slightly modifying the facts so that he has the bad cough and fever but no problem with his breathing. While he has pneumonia, since he has fluid in some alveoli, he has preserved lung function. Assume that his doctor properly diagnoses and treats Mr. Smith, and he gets better. But then it's not clear why his case of pneumonia counted as disease, according to Boorse's or Wakefield's theories, since it's not clear where the dysfunction was. The lungs were able to carry out their function of gas exchange, and the immune system carried out its function of fighting the infection. Although the cough and fever were unpleasant, they were also important components of the body's response to the microbe. People without a good cough are at greatly increased risk of respiratory infections, and fever may well be an adapted response that weakens the bacteria's ability to reproduce.

So although a serious case of pneumonia involves biological dysfunction, it's not clear that a mild case does too. And we can come up with many cases like this. Although measles is a disease, in most cases there is no clear dysfunction.

Boorse responds to similar concerns by pointing to what he calls dysfunction at the cellular level: even in a cold, he writes, cells in the mucosa are dying. Responding to some critics, he claims that the death of a single cell means that disease is present (1997, pp. 50–51). But this makes the necessary condition so easy to satisfy that it verges on triviality. During menstruation, after all, there is massive cell death as the uterine lining is shed. And during the third-trimester of pregnancy the large uterus interferes with the normal function of the bladder storing urine and of the veins carrying blood back from the legs. But menstruation and pregnancy are normal, healthy conditions.

Wakefield's 'value criterion' blocks some of the attacks on Boorse's theory based on such cases, as described above for homosexuality. But since the value criterion requires that there is 'some harm or deprivation of benefit to the person, as judged by the standards of the person's culture,' the theory faces some of the same counterexamples that confronted King (1954). If a culture values eczema or bound-feet, then these conditions will not count as pathological. Finally, even without the cultural relativity, making harm a necessary requirement opens the theory to counterexamples involving diseases that benefit their victims, such as flat feet keeping a young man out of the army or cowpox conferring immunity during a smallpox epidemic.

At this point, it starts to feel like attempting to craft a definition of disease is like the scene in the movie *Fantasia* where the sorcerer's apprentice is trying to eliminate the magical brooms: destroy one, and two are created. Or, it's like a nightmare version of Hans Brinker: plug one hole in the dike, and immediately another one opens up. At this point, it makes sense to reconsider the project overall.

5.5. CONCEPTUAL ANALYSIS AND AN ALTERNATIVE METHODOLOGY

5.5.1. Conceptual Analysis and Criteria of Application

All these theorists, and basically all contributors to the debate, have assumed that the goal is to provide an adequate *conceptual analysis* of disease. Current accounts of the project of conceptual analysis generally assume that the target is describing either (i) the meaning of the concept in question or (ii) the 'criteria of application' in the minds of competent speakers (cf. Neander, 1991). And, as mentioned above, both these approaches adopt the *classical view* of concepts, where concepts are represented by short lists of individually necessary and jointly sufficient conditions. Here I will argue that conceptual analysis – formulated in keeping with (i) or (ii), and the classical view of concepts associated with this project – is deeply problematic.

The claim to be uncovering criteria at work in the minds of speakers runs afoul of psychological research on the use of concepts. First of all, psychologists have undermined the classical view by showing that objects are usually classified based on characteristics that are not strictly necessary (Rosch and Mervis, 1975, pp. 382–439; Ramsey, 1992, pp. 59–70). For example, it counts towards an object's being a chair that it has four legs, but some chairs don't have legs at all. The fact that an animal flies counts towards its being a bird, according to the average speaker, but some animals are birds even though they don't fly.

Second, research has shown that the way speakers classify objects is a matter of degree, rather than an all-or-nothing affair, as the classical view implies (Ramsey, 1992, p. 63). Items can count as better or worse examples of a given concept. For instance, speakers see a kitchen-table chair as a better example of a chair than the five-foot-high chair that a tennis referee sits in, and speakers treat falcons and hawks as better examples of birds than ostriches. This ordering of items is reflected by many different measures: more central examples of a concept are learned more easily, come to mind more quickly, and influence thought more powerfully (ibid.). In addition, context can change which properties count as more important for classifying an object and for determining whether it is a good example of the concept in question (Ibid.).

These findings have led psychologists to adopt models that are closer to Wittgenstein's 'family resemblance' (1953) picture of concepts than to the classical view. According to some models, for instance, objects are classified based on whether they resemble a prototype in certain ways, measured according to some 'similarity metric.' In one quantitative model, a concept is represented by a list of properties, and each property has a point value: items that accrue enough points count as falling under the concept. In this way, different objects falling under the same concept may share few (or perhaps no) important properties. And it's relatively easy to make sense of which items count as a 'good example': a higher point total means a better example.

Neander (1991) describes her project of conceptual analysis – in this case, directed at the biological concept of function – as aiming to uncover the 'criteria of application' that are at work in the minds of biologists. And she argues that psychological research that undermines the classical view, described above, does not weaken her project since 'function' is a theoretical term used by experts, rather than one used by general speakers in common language. 'Function' for biologists, she claims, is like 'water' for chemists, with specific necessary and sufficient conditions for its use (1991, pp. 172–173).

I've argued elsewhere (Schwartz 2004, p. 146) that the analogy is not apt for 'function,' and it certainly isn't for 'disease' either. Although it is true in both cases that 'experts' – biologists and doctors use the term respectively – there is no commonly accepted technical definition that these experts share, as in the case of chemist's use of the term 'water.' The absence of any technical meaning for 'disease' can be seen, for instance, in the heated disagreements among experts over whether infertility, short-stature, or obesity are truly diseases (Konner, 1999, pp. 55–60). In fact, it is the lack of a commonly accepted meaning for 'disease' and for 'function' that attracts the interest of philosophers to these areas.

Another problem for conceptual analysis, independent of the findings that undermine the classical view of concepts, is the fact that people may rely on widely varying criteria when applying a given concept. For example, it is likely that different people will use the term 'disease' in different ways: there may be significant disparity between the use of the term by laypeople and by doctors, as well as among the various subfields of medicine. Even a single person may use the term in different ways at different times. There's really no way to characterize the range of variation unless psychologists were to study the use of the term extensively. And this is a project very different from the sort of analysis that philosophers generally employ. Either these philosophers have misdescribed the project, or they've chosen an unsuitable method for pursuing it.

5.5.2. Conceptual Analysis and Meaning

A different, and more common, defense of conceptual analysis avoids any claim concerning people's use of the term in question, and instead aims to uncover the term's or the concept's *meaning*. According to this picture, each concept has a meaning that lies behind the actual usage and common understanding of the concept and which guides its use in some way. Competent speakers may apply the concept by using similarity to prototypes or other methods, but the meaning, it is argued, is understood by all and can be discerned by careful philosophical study.

Although this picture is attractive in some ways, it assumes a notion of meaning that has been seriously undermined by analytic philosophy of language. One way to articulate the problem is as follows. Assume for a minute that such a meaning of 'disease' exists, and that we are assured that Mr. Smith has a disease. Then all other claims about Mr. Smith's condition can be classified as *analytic*, i.e. following from the meaning of 'disease,' or as *synthetic*, i.e. requiring more information. Which statements are analytic and which are synthetic depends on which definition of 'disease' is correct. For instance, if Culver and Gert are right, then it follows *analytically* that Mr. Smith is at increased risk of an evil; it follows just *synthetically* that his lungs are dysfunctioning. In contrast, if Boorse's account of 'disease' is correct, then the statement that his lungs are dysfunctioning follows analytically and the statement that he's at increased risk of an 'evil' follows synthetically.

The problem is that good arguments have undermined the claim that there is a fact of the matter about which statements are analytic and which are synthetic, given the truth of a certain proposition. Quine (1953) highlights the difficulty determining whether 'he is unmarried' follows analytically or synthetically from 'he is a bachelor,' and explains why, given certain assumptions, such a distinction cannot be a factual matter (1961). Putnam (1962) uses examples from history of science to show how problematic it is to determine which statements of science are analytic, in the sense of following just from the meaning of the terms, and which are synthetic. For example, he points out that the equation '$e = 1/2\,mv^2$' was initially a *definition* of kinetic energy and thus analytically true. But later arguments for relativity theory disproved the equation, suggesting that at some point it had become a synthetic claim. It is not 'happy,' Putnam says, to force all scientific statements into the categories of *analytic* or *synthetic*.

If this is so, then it also can't be the case that the terms of science have strict definitions, the sort that might be uncovered by conceptual analysis. I will not delve into the arguments for or against the analytic/ synthetic distinction further here, beyond pointing out that the important arguments attacking it undermine the idea that the analysis of 'disease' should be seen as a search for the meaning of that concept.

5.5.3. Conceptual Analysis, Natural Kinds, and Stipulation

In order to avoid such problems, some philosophers have suggested that the analysis of concepts should be seen as a search for necessary and sufficient conditions determining the extension. This picture seems to fit discoveries concerning natural kind terms, such as finding out that water is just H_2O. Millikan (1989) presents her definition of 'function' in this light, not as a claim about meaning, but as a characterization of a biological phenomenon. And she adamantly denies that her 'theoretical definition' has anything to do with an attempt to uncover the meaning of the term or the criteria used by speakers for applying it.

There's much to be said for this view, although its overall attractiveness depends on what account of natural kind terms it assumes. On some views, the criteria for membership in the extension of a natural kind term is *discovered*, as humans discovered that gold is all and only atoms with atomic number 79 and water is all and only molecules of H_2O. On this view, the terms 'gold' and 'water' always had these extensions, even if nobody knew it (Putnam, 1975). Opposing this view is the idea that the boundaries in such cases are *chosen*, in some sense, rather than discovered (Donnellan, 1983, pp. 84–104; Dupre, 1993).[1]

But whatever conclusion is drawn for 'gold' or 'water,' the case for 'disease' most likely best fits the latter model. As scientists have acquired better and better understanding of diseases and their causes, they find not a unifying microstructure, as for gold or water, but variation. While many have sought an *essence* that all and only diseases share, this quest has been blocked at every step by variability and heterogeneity. Any definition that would draw a sharp line through all conditions, determining for each whether it is a disease or not, looks like the imposition of a decision, rather than the application of a discovery.

This means that adopting any precise account will impose at least some changes on our currently non-reflective and relatively unprincipled way of distinguishing disease from health. Choosing a definition will partly involve deciding which changes from current practice are acceptable. This picture of the philosophical examination of the concept of disease resembles a general approach put forward in analytic philosophy by Carnap (1950, pp. 3–8) and Quine (1961, pp. 257–262).[2] As Quine writes,

We do not claim synonymy. We do not claim to make clear and explicit what the users of the unclear expression had unconsciously in mind all along. We do not expose hidden meanings, ...We fix on the particular functions of the unclear expression that make it worth troubling about, and then devise a substitute, clear and couched in terms to our liking, that fills those functions. (pp. 258–259)

While Quine's overall approach to language and meaning, with its strict behaviorism, does not fit current analytic philosophy, this recommendation is in keeping with the pragmatic bent of much current thought. In a previous paper (2004), concerning the concept of 'function,' I've recommended this approach and called it 'philosophical explication' (pp. 143–145).

5.6. PHILOSOPHICAL EXPLICATION OF 'DISEASE'

I believe that the philosophical study of 'disease' should similarly be framed as a project of philosophical explication, and thus should give up the attempt to *discover* the concept's true meaning, its criteria of application, or the boundaries of its extension. Instead, accounts should be seen as proposed *new* definitions, or groups of new definitions, that can do much of the same work that was done by the vaguely defined concept of disease in the past. Any definition will impose at least some changes and carry at least some counter-intuitive consequences. Imposing such precision may also mean that no single concept can be relied upon in all situations; there may need to be different definitions for different contexts.

From this perspective, the long debate between normativism and naturalism – including the discussion of the four theories described above – serves a crucial role by displaying the advantages and disadvantages of the various approaches. Choosing an account becomes not so much a hunch about which theory is correct but instead a choice of which theory to clarify and apply.

One reasonable question at this point is to ask whether definitions should be stated as necessary and sufficient criteria, as in the classical view, or instead in keeping with some other model, such as the psychologists' prototypes and similarity metrics. The attraction of the classical view is the possibility of providing clear answers in debates over which conditions are really diseases. Many philosophers and physicians attempt to define 'disease' partly as a way to decide how to classify and whether to treat borderline or complex conditions such as obesity, infertility, or short stature (Konner, 1999; Schwartz, 2001). However, once we see the acceptance of a given definition as a decision rather than a discovery, this motivation for defining 'disease' is greatly diminished. Using prototypes and similarity metrics has the advantage of being more in keeping with the actual use of 'disease' and other concepts.

In conclusion, I would like to return to the dysfunction-requiring approach, to evaluate how well it does if we adopt philosophical explication as our project and use prototypes and similarity metrics to represent the concept. In particular, the DR approach does a good job making sense of prototypical cases of disease and distinguishing them from healthy, undesirable conditions. Start with the following three types of prototypical diseases:[3]

1) Infection: severe bacterial infections such as pneumonia, sepsis (infection in the blood), or dysentery (infectious diarrhea);
2) Injury: broken bones or other sorts of wounds (e.g. puncture wounds or lacerations);
3) Organ failure: kidney or liver failure, blindness, or deafness.

I believe that within each category, the prototypical cases are the most serious ones, i.e. where there is a raging infection, a serious injury, or complete organ failure. And in all these cases, there is clear biological dysfunction as well. In (1) there is dysfunction of the lungs, the GI tract, etc. In (2) there is failure of the broken bone to provide support or of the lacerated skin to keep out infection. In (3) there is dysfunction of the kidney, liver, eyes, or ears.

These cases will also be classified as disease by a value-requiring, non-DR, approach, since each carries pain, loss of freedom, risk of death, etc. But these undesirable outcomes do not distinguish the conditions clearly from other undesirable but healthy conditions. It may be worse to be impoverished or illiterate than to have kidney failure, for example, given the availability of dialysis. Even being deaf may not impose as much disability or loss of freedom as does being unemployed or being swindled out of one's life savings. So if we focus on proto-typical cases, which I think makes sense to do as a first step, I believe that a DR criterion works better than a VR one.[4]

This approach still carries the unattractive implication that homosexuality may count as a disease, and this does not fit my intuitions or those of most of the current medical and psychiatric world. The inclusion of a harm requirement, as in Wakefield's account, answers that problem, but such a requirement brings problems of its own, as described above. It may be that this is a place where a sharpened definition of 'disease' carries an unavoidable counter-intuitive consequence.

The DR account also faces significant challenges as it moves from severe diseases to less serious ones. As mentioned above, minor injuries and early infections do not clearly involve any biological dysfunction, such as in the case of early pneumonia or most cases of the measles (Section 5.4.3). A DR account of the sort I'm sketching might respond by propounding some sort of similarity-metric intended to link such conditions to the more serious ones. Measles and other viral infections, and bronchitis or early pneumonia, are all processes that carry certain risks of dysfunction, and thus risks of causing prototypical disease.

I believe that there is promise here for an interesting and fruitful account of disease. At the same time, there may be equally promising accounts based on a value-requirement, perhaps also stated in terms of prototypes and similarity metrics. Once we accept that the project of defining 'disease' is a constructive one, and that definitions do not have to be stated according to the classical view, there is room for many approaches. Where there was once stalemate, let there be variability and free choice.[5]

Indiana University Center for Bioethics, Indianapolis, Indiana, USA

NOTES

[1] See Ebbs 2000, 2002 for a more complete analysis of this question.

[2] Gary Ebbs has recently utilized a similar project in philosophy of language (2002), and he first recommended the Quinean approach to me (personal communication). Ramsey (1992, 69) also mentions Carnap's (1950) approach as a possible response to problems with conceptual analysis. See also Schwartz (2004), and Lewens (2004).

[3] Of course, I make no claim that these are all the prototypes or that equally good but incompatible ones could be chosen.

[4] Sadegh-Zadeh (2000), brought to my attention after I wrote this paper, also proposes focusing on prototypical cases of disease first in analyzing the concept.

[5] Thanks to Christopher Boorse, Art Caplan, Gary Ebbs, Gary Hatfield, Eric Meslin, and Fred Tauber for helpful discussions relating to this work. Also thanks to the organizers, Harold Kincaid and Jennifer McKitrick, and audience of the conference on 'Philosophical Issues in the Biomedical Sciences,' University of Alabama, Birmingham, May 15, 2004.

REFERENCES

Boorse C (1977) Health as a theoretical concept. Philos Sci 44:542–573

—— (1987) Concepts of health. In: D Van De Veer and T Regan (eds) Health care ethics: an introduction. Philadelphia, PA, Temple UP, pp 359–393

—— (1997) A rebuttal on health. In: Humber JM, Almeder RF (eds) What is disease? Humana Press, Totowa, NJ, pp 1–134

Carnap R (1950) Logical foundations of probability, 2nd edn. Chicago University Press, Chicago, IL

Culver C, Gert B (1982) Philosophy in medicine: Conceptual and ethical issues in medicine and psychiatry. Oxford University Press, New York

Donnellan KS (1983) Kripke and Putnam on natural kind terms. In: Ginet C and Shoemaker S (eds) Knowledge and mind. Oxford University Press, Oxford, pp 84–104

Dupre J (1993) The disorder of things: Metaphysical foundations of the disunity of science. Harvard University Press, Cambridge, MA

Ebbs G (2000) The very idea of sameness of extension over time. American philosophical quarterly 37:245–268

—— (2002) Learning from others. Nous 36:525–549

Engelhardt H, Tristam, Jr (1974) The disease of masturbation: Values and the concept of disease. B Hist Med 48(2):234–248

Gert B, Culver C M, Danner-Clouser K (1997) Bioethics: A return to fundamentals. Oxford University Press, New York

King L (1954) What is disease? Philos Sci 21(3):193–203

Konner M J (1999) One pill makes you larger: The ethics of enhancement. American prospect, 42:55–60

Lewens T (2004) Organisms and artifacts: Design in nature and elsewhere. MIT Press, Cambridge, MA

Millikan R G (1989) In defense of proper functions. Philos Sci 56:288–302

Neander K (1991) Functions as selected effects: The conceptual analyst's defense. Philos Sci 58:168–184

Putnam H (1962) The analytic and the synthetic. In: Feigl H, Maxwell G (eds) Minnesota studies in the philosophy of science, vol. 3. University of Minnesota Press, Minneapolis

—— (1975) The meaning of 'meaning'. In: Mind, language, and reality: Philosophical papers, vol. 2. Cambridge Univ. Press, Cambridge, pp 215–271

Quine WVO (1953) Two dogmas of empiricism. In: From a logical point of view. Harvard University Press, Cambridge, MA, pp 20–46

—— (1961) Word and object. MIT Press, Cambridge

Ramsey W (1992) Prototypes and conceptual analysis. Topoi 11: pp 59–70

Reznek L (1987) The nature of disease. Routledge & Kegan Paul, London

Richman KA (2004) Ethics and the metaphysics of medicine: Reflections on health and beneficence. MIT Press, Cambridge, MA

Rosch E, Mervis C (1975) Family resemblance: studies in the internal structure of categories. Cognitive psychology 8:382–439

Rosenberg C (1992) Framing disease: Illness, society, and history. In: Rosenberg CE and Golden J (eds) Framing disease: Studies in cultural history. Rutgers University Press, New Brunswick NJ, pp xiii–xxvi

Sadegh-Zadeh K (2000) Fuzzy health, illness, and disease. J Med Philos 25(5):605–638

Schwartz PH (2001) Genetic breakthroughs and the limits of medicine: Short stature, growth hormone, and the idea of dysfunction. St. Thomas Law Review 13(4):965–978

—— (2004) An alternative to conceptual analysis in the function debate. The monist 87(1):136–153

Wakefield JC (1992a) The concept of mental disorder: On the boundary between biological facts and social values. Am Psychol 47(3):373–388

—— (1992b) Disorder as harmful dysfunction: a conceptual critique of DSM-III-R's definition of mental disorder. Psychol Rev 99(2):232–247

—— (1999a) Evolutionary versus prototype analyses of the concept of disorder. J Abnorm Psychol 108(3):374–399

—— (1999b) Mental disorder as a black box essentialist concept. J Abnorm Psychol 108(3):465–472

Wittgenstein L (1953) Philosophical investigations. Trans. Anscombe GEM. Macmillan Publishing Co., Inc, New York

6. RACE AND SCIENTIFIC REDUCTION

6.1. THE PROBLEM OF RACE IN CONTEMPORARY SCIENCE

The Human Genome Project has created some anxiety about the reemergence of 'race' in biology and medicine. The concept of race, it seemed, had been eliminated from physical anthropology and human biology during the twentieth century. Recent analyses of genetic variation, however, show a geographic pattern to genetic variation that roughly corresponds to traditional notions of race. Moreover, there are persistent correlations between race and health. American Blacks suffer higher rates of HIV infection, diabetes, hyper-tension, cardiovascular disease, and so on, than American Whites. In May, 2003, the Human Genome Center at Howard University convened a workshop to address the question 'What does the current body of scientific information say about the connections among race, ethnicity, genetics, and health?' (Patrinos, 2004). The papers were published in a November 2004 supplement to *Nature Genetics*. Participants agreed that while genetic diversity clusters geographically, the variation is not structured into races or sub-species. Some went on to argue that race remained useful as a proxy to identify genetic variation relevant to health. The conceptual frailty of this position was illustrated by the *New York Times* report on the conference results. Under the headline 'Articles Highlight Different Views on Genetic Basis of Race,' the article began: 'A difference of opinion about the genetic basis of race has emerged between scientists. ...' (Wade, 2004, A18). The reporter thus translated disagreement about the latter issue (whether race is a useful proxy) into disagreement about the former (whether race has a genetic basis). The confusion in *The New York Times* is not mere sloppy reporting. There is a tension between the thesis that there are, genetically speaking, no human races and the thesis that race might be a useful proxy for identifying medically relevant genetic differences. How could race be useful as a proxy if race had no genetic basis? In spite of protests to the contrary, has the Human Genome Project really shown that race is real after all?

A brief review of the history of the race concept and of recent work in genetics will help clarify these questions. As a scientific concept, race arose in the eighteenth century and was substantially developed during the nineteenth. In his *Systema Naturae (1758/1964)*, Linnaeus divided the species *Homo sapiens* into four sub-species, or races. They were distinguished by geographic and somatic features. Their names identify the races by continent: *Americanus, Europaeus, Asiaticus*, and *Afer*. It is noteworthy that the somatic features used to differentiate the races include both skin color and the dominance of one of the four humors. Native Americans are thus choleric, Europeans sanguine, Asians melancholy, and Africans phlegmatic. Eighteenth century medical thought was dominated by the Galenic system where

H. Kincaid and J. McKitrick (eds.), Establishing Medical Reality, 65–82.

humoral balance was used to explain health and illness. A relationship between race and health was thus established at the inception of racial theorizing.

In the nineteenth century, accounting for the difference among human groups was an important anthropological problem. Two centuries of exploration and trade had made human diversity vivid to European theorists. Moreover, the differences apparently clustered in a coherent way. Skin color and other aspects of somatic form seemed to vary with differences in language, technological development, social organization, and cultural forms. Health differences among the races maintained their significance, even as Galenic medicine gave way to germ theory. Black Americans were thought to be especially susceptible to tuberculosis, syphilis, and other diseases, while being resistant to malaria and yellow fever. Some physicians attributed the difference primarily to physiological differences between blacks and whites (Cartwright, 1871, pp. 421–427). Even those who recognized poor sanitation or diet as important risk factors typically attributed differences in sanitation or diet to purported racial characteristics such as 'indolence' (Harris, 1903, pp. 834–838) or 'fatalism' (Folkes, 1910, pp. 1246–1247). The human species thus seemed to divide into sub-species with systematically distinct properties. Race was taken to be a property of humans – Blumenbach called it the *nisus formatives* – that accounted for our somatic, medical, psychological, social, cultural, and linguistic differences. In the nineteenth century, 'race' was a powerful explanatory posit.

While racial theorizing dominated nineteenth century thinking about humans, there was opposition from within anthropology. Franz Boas was one of the first to make empirical arguments against race, and by the early twentieth century, his voice had been joined by a chorus of others. The main target of Boas' investigation was the coherence of racial classification. This was important because the identification of a sub-species requires a set of traits that are 'concordant.' This means that variation in one of the traits reflects variation in the others (Livingstone, 1962, pp. 279–281). For example, a mixed bucket of tennis balls, golf balls, and baseballs divides naturally into sub-groups because size, surface, and weight are concordant traits. Separating the balls by size yields the same sub-groups as partitioning by surface texture or weight. Surveying known cultures and languages, Boas showed that somatic form, language, and culture are not concordant. People with different body types may speak the same language (Black and White Americans speak English); persons with the same somatic form may speak unrelated languages (Finnish and Swedish are in different language families). Similarly, culture and language vary independently (Boas, 1894/1974; 1911). Boas concluded that the three criteria (language, culture, and somatic form) together could not provide a consistent racial classification. This conclusion had two important ramifications. First, it began to rob the concept of race of its explanatory power. If humans did not exhibit systematic variation of somatic form, language, and culture, then race could not explain such differences. Second, it shifted the focus of the debate to human biology and physical anthropology. If concordant traits were to be found, they would have to be physical.

Early twentieth century research in physical anthropology continued to undermine the viability of racial theorizing. By studying changes in the children of immigrants, Boas showed that many of the physiological features used to classify race – head size, stature, etc. – depended on diet or other environmental factors. He also found that variation within racial groups on such measures was greater than the variation between them. By 1936, Julian Huxley and A. C. Haddon were able to argue that physical criteria for race were not concordant. Classifying people by blood type yields a different partitioning than a classification by skin tone, which is different from a classification by hair texture, and so on. Moreover, Huxley and Haddon used the new science of genetics to argue that one would not expect concordant traits among populations of a species, like *Homo sapiens*, that underwent continual interbreeding. The lack of concordance among the physiological traits used to identify races was a severe blow to the explanatory power of the race concept. Human variation just was not systematic enough to form distinct sub-groups. Hence, any attempt to explain human variation by appealing to a hidden *nisus formatives* or by descent from an original pure racial stock was bound to fail. Haddon and Huxley concluded: 'These considerations rob the terms *race* or *sub-species*, as applied to existing human groups, of any significance' (Huxley and Haddon, 1936, p. 219).

Contributions to the Human Genome Center Workshop confirmed Huxley and Haddon's conclusion. Compared with other species, human beings are remarkably homogeneous. For example, biologists generally recognize three subspecies of the common chimpanzee, *Pan Troglodytes*. Genetic variation within one of the Chimpanzee sub-species is roughly twice the genetic variation found within the whole human species (Fisher et al., 2004, pp. 799–808). Moreover, what little genetic variation there is within the human population does not cluster in a way that would constitute sub-species (Keita et al., 2004). In the sense of the word meaningful to biologists, there simply are no human races.

On the other hand, geographic variation is an important aspect of the concept of race, both in biology and in nineteenth century anthropology. Contemporary genetics has found that what little variation there is among modern humans roughly corresponds to the continental origin of their ancestors (Jorde and Wooding, 2004, S17–S20). Does this show that human races exist? In their contribution to the Workshop, Jorde and Wooding answered in the negative. Their analysis measured genetic similarity with 100 *Alu* insertion polymorphisms and 60 short tandem repeat (STR) polymorphisms. The resulting clusters of similar genotypes roughly corresponded to the traditional geographical division of races. Jorde and Wooding argue that this result must be understood in the context of two further pieces of evidence. First, the variation among individuals was virtually continuous. So, a comparison of, say, a person from East Asia with another from Europe would show that each individual is more similar to individuals from the same geographic region than she is to those from the other region. But if the comparison included individuals from a region in between, like the Indian subcontinent, it would show substantial overlap with both European and Asian populations. While genetic differences reflect geographic separation and the history of migration, continual interbreeding has prevented sub-species boundaries from developing. Second, Jorde and Wooding's data confirms that the vast majority of human genetic variation occurs *within* continental populations. For the 100 *Alu* insertion polymorphisms, 86% of the variation occurs between individuals and within continents; for the STR polymorphisms, 90% occurred within continents. Practically speaking, this means that for many significant genetic variants, continent of origin will be a very poor indicator of whether an individual possesses the variant. Jorde and Wooding conclude:

> The picture that begins to emerge from this and other analyses of human genetic variation is that variation tends to be geographically structured, such that most individuals from the same geographic region will be more similar to one another than to individuals from a distant region. Because of a history of extensive migration and gene flow, however, human genetic variation tends to be distributed in a continuous fashion and seldom has marked geographic discontinuities. Thus, populations are never 'pure' in a genetic sense, and definite boundaries between individuals or populations (e.g., 'races') will be necessarily somewhat inaccurate and arbitrary. (Jorde and Wooding, 2004)

As Haddon and Huxley concluded almost 70 years earlier, 'race' has little or no significance in human biology.

While twentieth century physical anthropologists were busy eliminating the concept of race, social scientists and health researchers continued to use it to frame their research. They discovered apparently robust correlations between race and a variety of social and medical variables. Moreover, while genetic variation among human groups is small, some of that variation has medical significance. This situation raises the question of whether the social and medical sciences should continue to use a concept of race when that concept has no biological basis.

The contributors to the Workshop phrased the question as whether race might be useful as a 'proxy' identifier of genetic difference so long as a more specific, individualized genetic test is lacking. The question is complicated, and none of the authors who published their opinions in the *Nature Genetics* Supplement argued for a definitive answer. While genetic variation clusters by geographic region, and thus roughly corresponds to 'race' as it is socially identified, there is substantial heterogeneity within any such group and overlap among groups. Clinically, this means that using socially identified race to indicate health risks can lead to tragic mistakes (Jorde and Wooding, 2004, pp. S28–S33; Rotimi, 2004, pp. S43–S47). The heterogeneity of racially identified groups can also lead to bias in epidemiological or clinical studies, if the relevant genetic variation is carried mostly by a particular sub-group (Tishkoff and Kidd, 2004). On the other hand, race is useful as a rough indicator of genetic risk factors in the same way that age or gender is (Jorde and Wooding, 2004; Tishkoff and Kidd, 2004). Moreover, if the concept of race were eliminated from medicine or public health, it would become impossible to identify injustices in the distribution of medical resources or environmental risk factors associated with racial discrimination.

Another facet of the question of whether race might serve as a proxy was not discussed by the contributors to *Nature Genetics*. What could it mean for 'race' to serve as a proxy? If we are to use one thing as a proxy for another, both parties to the relationship need to be well defined. We have a good grip on genetic variation, but what is 'race'? By what criteria is it to be identified? Is it possible to have a coherent and useful notion of race in the absence of concordant traits or systematic genetic variation? The brief history of racial theorizing sketched above is sufficient to demonstrate that these questions cannot be taken for granted. Before we can raise the issue of whether race might be useful as a proxy for genetic variation, we need to be clear on what the concept of race should be in medicine and the social sciences. In particular, how is 'race' to be conceptualized in medicine and the social sciences, given that there is no genetic or biological role for the concept?

The philosophical issue looming in the background is the question of how to understand conceptual change in the sciences. As scientific theories develop and replace one another, concepts emerge, change, and disappear. There are two specific questions about this process to which we must attend. The first question concerns conceptual content. In virtue of what does a scientific concept mean what it does? Some kind of answer to this question must be available if we are to identify conceptual change at all. For example, Galenic medicine explained disease phenomena (as well as psychological disorders and personality traits) in terms of the balance among four bodily fluids: blood, phlegm, black bile, and yellow bile. In Galenic medicine, blood was hot, and a superfluity of blood produced fevers. When pathogens and lesions replaced the humors in explanations of illness, physicians still talked about 'blood', but the concept had apparently undergone a dramatic change. What changed, and why think there was a change at all? In the latter part of this essay, we will ask similar questions about 'race.' Is the twenty-first century concept of race the same concept as was used in the nineteenth? We will presuppose a view of conceptual content, but it is beyond the scope of this essay to inquire deeply into these matters. This essay will adopt a broadly Sellarsian notion of conceptual content (Sellars, 1963). Two concepts have the same meaning insofar as they have the same explanatory role, that is, they explain the same phenomena and the same inferences can be made with them. In other words, the distinctive content of a concept is the difference it makes to the theory. The significance of the concept of race, then, is determined by the explanations and inferences in which it plays a role. As these explanatory roles change, the meaning or content of the concept changes.

The second specific question about conceptual change in the sciences concerns reductionism. Literature on this topic in the last 50 years or so has shown reductionism to be a multi-dimensional affair. For example, nineteenth century theories of electromagnetism postulated the existence of a medium, the ether, through which electromagnetic waves were propagated. The work of Lorentz and Einstein showed how space and time could be reconceptualized to make the 'ether' superfluous. The concept of ether disappeared entirely from physics. It was *eliminated,* and no concept fulfils a similar theoretical function. The concepts of space and time were not eliminated in the same sense. They were *replaced* in relativistic physics by the concept of spacetime, which plays a role similar to that played by 'space' and 'time'. Because the elimination or replacement of concepts occurs within a scientific domain (physics), it has been called 'intra-level' reduction by Robert McCauley

(1984). A different set of questions about reduction arise when we think about theories arranged hierarchically, with medicine or psychology on a higher level than biology, and biology higher than chemistry. 'Inter-level reduction' is at issue when we wonder whether social institutions can be reduced to the actions of rational individual agents, or whether beliefs can be reduced to brain states (McCauley, 1984). Inter-level reduction has been sought by philosophers and scientists for two reasons: epistemic vindication and ontological economy. When it can be shown that the explanatory force of higher-level concepts or propositions can be exhausted (at least in principle) by an epistemically more secure lower-level theory, the higher-level theory gains support. In such a case, the concepts would not be eliminated from the higher-level theory; rather, they would be identified with the concepts of the lower-level theory. Such a reduction would also achieve ontological economy because the higher-level concepts would carry no new ontological commitments.

McCauley's distinction between inter-level and intra-level reduction sharpens the issue about race. In the nineteenth century, race played a role across the hierarchy of the sciences. Racial difference explained human variation in phenotype, psychology, language, culture, and disease susceptibility. Because the traits used to distinguish the races were not concordant, the concept of race underwent an intra-level reduction in human biology and physical anthropology. Insofar as 'race' no longer has an explanatory role at this level, we can say that the concept has been eliminated. Robust correlations between race and health, as well as between race and social variables, meant that the concept of race did not undergo an intra-level reduction in medicine and sociology. This left a concept of race operating in the higher level science without a correlate in the lower level science. If inter-level reduction is an important goal of science, then a higher-level concept without a lower-level correlate concept would seem problematic. The uncertainty about using race as a proxy for genetic difference thus reflects a deeper philosophical question. Does the intra-level reduction (elimination) of race at the lower, biological level force an intra-level reduction (elimination) of race in medicine or the social sciences?

The rise of non-foundational epistemologies and continued research into the complexities of inter-theoretic relations has led many philosophers of science to reject the goal of inter-theoretic reduction. Post-positivist philosophers like Richard Miller (1987) and John Dupré (1993) have argued that scientific disciplines are both epistemically and ontologically autonomous. They are epistemically autonomous in the sense that the theories of a particular discipline can be fully supported by observations and methods that do not reduce to theories at a lower level. Inter-theoretic reduction fails, but this does not undermine the epistemic status of the unreduced theory. Dupré and Miller also argue that unreduced theories have autonomous ontological commitments. Dupré dubbed this position 'promiscuous realism' (1993, p. 7). Several recent writers on race, including Michael Root (2000), Ron Sundstrom (2002a, b), and Richard Miller (2000) have defended promiscuous realism about race. They argue that race has an ineliminable explanatory role in medicine and the social sciences. As a result, these disciplines are committed to the existence of race. Since ontological commitments are autonomous, this commitment is not undermined by the elimination of race in physical anthropology.

Promiscuous realism about race permits a solution to the puzzle about what to make of racial generalizations in medicine, when there is no such thing as race in human biology. The distinctions among races cause people to be treated differently and to have different life opportunities. Some of these differences are relevant to health. Racial health differences, or at least some of them, are thus explained by the social reality of race. Notice that this position is distinct from the relatively weak claim that race is a good proxy for genetic differences. The promiscuous realists argue for the stronger claim that as a social kind, race directly explains health outcomes. From this stronger position, promiscuous realism can answer the question about what it means for race to be a proxy. Treating race as real social kind, in the way that the promiscuous realists do, provides a robust conception of what race *is*. Hence, we can raise and answer the question of whether there is a correlation between genetic variation and the races of a particular time and place. The 'time and place' qualification is essential. Races arise out of treating people in a particular way, and this varies significantly with location and historical period. Assuming that the correlations between race and genetic risk factors are robust enough to be useful, the promiscuous realist's position supports the use of race as a proxy.

Promiscuous realism about race thus provides a tidy answer to the cluster of questions we are exploring. By adopting an anti-reductionist ontological stance, it provides an account of 'race' that is not troubled by the intra-level reduction of 'race' in physical anthropology. By treating race a real social kind, it explains the significance of medical and social scientific generalizations about race. Finally, it provides an account of the conditions under which race might be a useful proxy for genetic variation in the diagnosis and treatment of disease. Before we acquiesce in this happy solution, two further issues need to be explored. First, we need to look more closely at the explanatory role of 'race' in medicine and the social sciences. What kind of explanations are these? Why do they not collapse without a biological basis for race? Second, we will return to the issue of conceptual change raised above. Is the concept of race as articulated by the promiscuous realists the same concept as was used by nineteenth century racial theorists? If it has changed significantly, to what extent is promiscuous realism a kind of realism about *race* at all?

6.2. RACE AND INTER-THEORETIC REDUCTION

The promiscuous realists want to hold that race is a social kind, and that commitment to the existence of this social kind is supported by its ineliminable role in social and medical explanations. To evaluate their position, we need to understand the precise role of race in social and medical explanations. We will proceed on two weak assumptions: (1) that being an answer to a why-question is a necessary condition for being an explanation, and (2) that explanatory autonomy is a necessary condition for ontological autonomy. These assumptions are weak enough to be shared by proponents and critics of promiscuous realism, yet they are robust enough to permit a substantial analysis of the explanatory autonomy of 'race.'

The erotetic model is a well-understood analysis of explanation that treats explanations as answers to why-questions (Garfinkel, 1981; Lipton, 1991; van Fraassen, 1980; Risjord, 2000).[1] The formal machinery of the erotetic model will be useful in our analysis because it permits a fine-grained understanding of the different roles of concepts and propositions that appear in explanations. According to the erotetic model, an explanation is an answer to a why-question, and why-questions have the form 'Why P, rather than $\{Q, R, \ldots\}$?' P is the *topic* of the why-question and the set of propositions, $\{Q, R, ..\}$, are the *foils*. The answer to a why-question must discriminate between topic and foils, and the *relevance criterion* specifies how such discrimination is to be made. The topic, foils, and relevance criterion entail a set of presuppositions for the why-question (van Fraassen, 1980, pp. 141–146; Risjord, 2000, pp. 72–79). These must be satisfied if the question is to be appropriately asked. For example, the topic of a why-question is presupposed to be true, and the foils are presupposed to be false. It makes no sense to ask 'Why are the lights on (rather than off)?' when the lights are, in fact, off. While a full treatment of presuppositions is impossible here, it has been argued that the why-questions of different domains have distinct presuppositions. This has important consequences for the issue of inter-theoretic reduction. The erotetic model lets us specify the conditions for reduction quite precisely, and it lends itself to a general argument that explanatory, inter-level reduction is not always possible (Garfinkel, 1981; Risjord, 2000).

Let us turn to the explanations on which the promiscuous realists rely. Root and Sundstrom emphasize the explanatory power of racial norms. Root writes that norms 'have great explanatory power, for a social scientist can explain why blacks are underrepresented in some trades or professions by citing a past or present rule or regulation which says to keep blacks out' (Root, 2000, p. S633). The form of this explanation is relatively clear. The topic is a population-level phenomenon, a distribution of occupation or disease. Why, we ask, do people identified as 'Black' in the United States have a higher blood pressure than those identified as 'White', rather than having the same blood pressure? The explanation is that there is a rule which recommends differential treatment of Blacks and Whites. The differential treatment then causes a difference in average blood pressure between Blacks and Whites. Suppose this sort of explanation were shown to be autonomous. Would it support promiscuous realism about race? It does not seem to do so. The promiscuous realists hold that race is a real social kind. In the above explanation, however, it is not any social kinds, but the rules, that do the explanatory work. Norms and rules make distinctions among people, and they sometimes thereby create or presuppose social kinds, but they do not always do so. For example, imagine a rule saying that no person who is less than four feet tall, is pregnant, or has a pacemaker may ride the rollercoaster. The existence of such a rule does not make 'being less than four feet, being pregnant, or having a pacemaker' a social kind. The ineliminability of norms in social or medical explanations would show only that discriminatory norms, not race, have an explanatory role.

A related kind of explanation is suggested by Sundstrom when he observes that 'Discrimination has been linked to high levels of blood pressure, stress, anger, and emotional and psychological distress in communities of color' (Sundstrom, 2002a, p. 97). Sundstrom here refers to a number of epidemiological studies that correlate discrimination with negative health outcomes. These studies try to control for socio-economic variables and all of the known risk factors in order to isolate exposure to discrimination as the independent variable. They find that individuals who take themselves to have been subject to discrimination have higher blood pressure (Krieger and Sidney, 1996, pp. 1370–1378), lower birth rate (Collins et al., 2000), etc. If we take this to be an explanation, then the why-question must be asking about the difference between White health outcomes and Black health outcomes, e.g. Why do Blacks have higher blood pressure than Whites, rather than the same? Unlike the previous example, the answer is not a norm or rule. Krieger, Collins, and their colleagues suggest that perceiving oneself to be the subject of discrimination explains the difference in health outcome. Of course, these perceptions may well reflect actual discrimination, and such discrimination may be sanctioned by the larger White community. But these background facts are not part of the explanation. The explanation would succeed even if the perceptions were largely mistaken, since it is the perception alone that is supposed to explain the difference. Being Black or White plays no role in the answer to these questions, so we cannot conclude that race has any explanatory power of its own.

Identity has become a popular area of inquiry for the social sciences. Miller emphasizes the role of race in the formation and maintenance of racial identities in his discussion of the explanatory power of race. Thinking of oneself as Black is, in particular times and places, a grounds for identifying with a community, adopting its goals as one's own and caring for its members (Miller, 2000, p. S649). Again, the form of this explanation is relatively clear. The topic is the actions and decisions of an individual, and the question is why she did this, rather than something else. The answer refers to the person's belief that she is a member of the community, and perhaps the value she puts on being a member. Explanations of this sort are, no doubt, important in the social sciences. But once again they do not give any interesting role to race; the explanation relies entirely on beliefs and values.

The three examples canvassed so far have been disappointing. The concept of race appears, but race as a social kind does little or no explanatory work. All of the heavily lifting is done by something else: norms, perceptions, beliefs, or values. The mention of 'race' only specifies these as norms of racial discrimination, perceptions of racial discrimination, or beliefs about racial identity. Having a race, even a socially identified race, plays no explanatory role at all.

The discussion so far has been unfair to the promiscuous realists insofar as it has taken explanations in terms of norms, perceptions, and identity to be independent. A more promising way to understand the promiscuous realists is to take norms, perceptions, and identity to be part of a more complex whole. Developing some of

Root's ideas, Sundstrom contends that a constellation of three social forces causes race to become real at particular times and places (Sundstrom, 2002a). First, racial categories need to be deployed. The concept of race needs to be used by members of a society to make distinctions among persons. Sundstrom and Root agree that not every social distinction creates a social kind. Two further features must be present. Individuals so divided need to think of themselves as Black, Hispanic, Asian, or White, and they need to act on that basis. Race thus needs to become a part of their self-identity and their grounds for identification with others. Finally, there need to be norms that depend on the racial category. These norms might be laws, like Jim Crow statutes, or they may be the more subtle and pernicious norms of racial discrimination. The norms need to distinguish among persons based on race and then prescribe differential treatment of some kind. The three criteria of division, uptake, and normativity define a particular sort of social kind, what we might call a *social status*. Having a social status is a product of being treated in a distinctive way and making that distinctive treatment part of ones' own reason for acting. Moreover, statuses are normative, with implications for ways in which the holder ought (not) to act and how others ought (not) to act with respect to them.

The three-criterion model of the social status of race permits us to distinguish between those norms and self-ascriptions of identity that create social statuses and those that do not. 'Race is like crime,' Root writes, but unlike postal code (Root, 2000, pp. S630–S631). By this he means that while we divide ourselves by postal code, this does not say anything about us. Zip codes divide us into categories, and Post Office rules prescribe differential treatment, but division created is not undertaken by those so divided. Being a felon, on the other hand, is a real social status according to the model.

It is easy to find explanations where behavior – either individual acts or group patterns of action – are explained by social statuses. Root's example of discrimination in occupation is best understood, not as an explanation by norms alone, but as an explanation by social status. Blacks are underrepresented in certain trades because they are *Blacks*. Being Black involves differential treatment and uptake of identity. The norm makes it difficult to enter the profession, and the undertaking of the identity means that individuals will be disinclined to try. Sundstrom cites Massey and Denton's *American Apartheid* (1993), where they take race to explain segregated patterns of housing in American cities. Again, it is not just the discriminatory norms that explain the housing pattern, but the status of being Black. Unlike the foregoing examples, these explanations give a robust role to race as a social kind. The fact that the individuals share a social status explains differential outcomes in health, housing, employment, and so on. The Root–Sundstrom analysis of race as a social status involving distinction, uptake, and norms is thus crucial to their promiscuous realism about race. Having a race plays an explanatory role only if we take race to be a social status of this kind.

In the conception of race as a social status, we have discovered a robust explanatory role for race in medicine and the social sciences. Now we must turn to the question of whether this explanatory role is autonomous. That is, does talk of 'race' in such explanations do nothing more than abbreviate descriptions at a lower level? For instance, can appeal to the social role of race be replaced by appeal to the beliefs of individuals? This is the question of inter-level reduction.

Explanatory, inter-level reduction would occur when questions raised at one level could be answered by reference to only the entities of the lower level and their properties. Where the higher-level theory identifies a pattern, the reductionist explanation would explain the pattern as a product of only the lower level entities. The explanation of a higher-level pattern in terms of lower-level entities presupposes that the existing pattern is one out of all of the possible combinations of lower-level entities, and that all the combinations are possible. That is, there are no constraints on the lower-level entities that must be described in higher-level terms. The presupposition of the why-question is then satisfied, and the inter-level reduction is successful (Garfinkel, 1981; Risjord, 2000). In other cases, however, the presupposition of the reductionist why-question will not be satisfied and the why-question cannot be answered. In such cases, there are higher-level relationships that restrict the interactions among the lower-level individuals. Where the interactions are restricted, not every logically possible combination of individual states is a real possibility for the system. The explanation must then appeal to the relationship among the lower-level entities to explain the higher level pattern. The explanation can not appeal only to the lower-level entities, and there would be no inter-level reduction.

Consider an example where the social and the psychological levels are in play. Suppose we notice that on a particular bus in Atlanta the Blacks are riding in the back and the Whites are in the front. Imagine two different time-frames for this event: 1954 and 2004. If we ask 'Why are the Blacks riding in the back and not the front of the bus (rather than both back and front)?' in 2004, the answer might be given entirely in terms of the beliefs, desires, and habits of the individuals. (This assumes, perhaps counterfactually, that in 2004 no discriminatory norms explain the segregation.) In this case, the why-question about a social-level pattern (the pattern of seating on this bus) is answered entirely in psychological terms. Here we have the beginning of a reduction of a social phenomenon (the seating pattern) to psychological phenomena (the attitudes of individuals). In 1954, however, matters are different. We cannot explain the seating pattern simply by referring to the motives of the individuals. While each person chose his or her seat, it would be false to say, baldly, that the Blacks are sitting in the back because they chose to. There is a law in force that commands a seating distribution of this kind. This sociological question about the seating pattern cannot be answered strictly in psychological terms. The law limits the possible seating patterns, and an adequate explanation must take it into account. Hence there is no reduction when we explain the 1954 seating patterns.

These arguments show that, in the right context, some questions about patterns of behavior cannot be answered by appeal to individual psychology alone. The explanation must also refer to norms and social statuses. In general, norms and the social statuses that depend on them have explanatory autonomy. The explanations identified by the promiscuous realists are an instance of this general pattern. The explanatory role of 'race,' when construed as a social status, depends on the existence of norms and patterns of uptake and response. Explanations that appeal to 'race' (again, construed as a social status) are therefore autonomous with respect to explanations that only appeal to properties of individual persons. There is no inter-theoretic reduction of race, at least insofar as this requires explanatory reduction. These are important conclusions because they vindicate the promiscuous realist position about race, at least partially.[2] Having secured this result, we may turn to the second group of questions that ended Section 6.1. What is the content of concept of race we have shown to be autonomous, and how is it related to the nineteenth century concept of race? Is the intra-theoretic elimination of race in biology consistent with the failure of inter-theoretic reduction of 'race' in contemporary, medical and social scientific explanations?

6.3. RACE AND INTRA-THEORETIC REDUCTION

The difference between the nineteenth and the twenty-first century conceptions of race should be quite clear from the foregoing discussion. The concept of race as used in contemporary medicine and social science is that racial identities create communities of persons who share goals and care about each other as members of the same race. Norms prescribe and proscribe behavior of those who have the race. The norms may be accepted as a positive expression of what it means to be a member of the community, or they may be challenged as unjust, oppressive, or stereotypical. The difference between this concept of race as a social status and the concept of race used in the nineteenth century is striking. Most importantly, race is no longer conceptualized a natural fact about a person. In the nineteenth century, 'race' was like eye-color; one had it independently of any social status. (Indeed, it is in virtue of this that race could be used as the explanation of and justification for the social status of persons of color.) In the twenty-first century race is not taken to be a natural property of persons. It is nothing more (or less) than a product of how people are distinguished and treated and how those distinctions are internalized.

The change in the concept of race from the nineteenth to the twenty-first century is so profound that it constitutes an intra-theoretic elimination. While we still use the same word, the nineteenth century concept of race has been replaced. The twenty-first century concept of race has different implications, a different explanatory role, and it divides the social world in different ways. The idea that race is a social status thus represents the culmination of a line of conceptual development that to which Boas, Haddon, Huxley, Montagu, Livingstone, and many others contributed. In 1936, Haddon and Huxley were ready to draw the same conclusion:

In most cases it is impossible to speak of the existing population of any region as belonging to a definite 'race,' since as a result of migration and crossing it includes many types and their various combinations. For existing populations, the word *race* should be banished, and the descriptive and noncommittal term *ethnic group* should be substituted. (Huxley and Haddon, 1936, p. 220)

The change in terminology from 'race' to 'ethnic group' reflects an important conceptual shift from the nineteenth to the twentieth century; it is not motivated by mere political correctness. This change was forced by two developments discussed in the first section of this essay. When race was eliminated from physical anthropology, it could no longer be taken as a natural fact about humans. All that is left of the concept is its social elements. Second, ethnicity and the concept of culture were developed so that they would explain human differences without appeal to biological races. This essay began with the question of whether the intra-level reduction of race in physical anthropology forced an inter-level reduction of race. We can now see that matters are more complicated. The elimination of race in physical anthropology forced a dramatic change in the concept of race as it was used in medicine and the social sciences. As a result of this change, the concept of race as a natural fact about humans has been replaced with the concept of race as a social status. Huxley and Haddon recommend marking this new conception with a new word: ethnicity.

It may seem like a confusion to conflate race and ethnicity. Sundstrom argues explicitly against it:

> To use ethnicity, culture, or class to understand US history or various social indicators would be to miss key features of social organization in the USA. Blacks were not enslaved because they were simply from the African continent, inhabited a lower socioeconomic class, or were thought 'primitives' ...; no, West African blacks were enslaved and kept in slavery because the were members of the black race, and were judged to be moral and intellectual inferiors who were not deserving of human rights. (Sundstrom, 2002a, p. 100)

Sundstrom's premise that United States history and social organization could not be understood without reference to race is surely correct. Race and ethnicity, however, are both social statuses constituted by differentiation, uptake, and norms. What is the difference, if any?

Since a social status is constituted by differentiation, uptake, and norms, *differences* in differentiation, uptake, and norms must create differences in status. The constitutive differences among statuses may be divided into two kinds: differentiation criteria and behavioral expectations. Differentiation criteria are used to identify those who have (or are eligible for) the status. For example, to qualify for the role of a married person in the United States, one must not already be married. To qualify for the role of a police officer, one must complete the right sort of training. These are part of what distinguishes the social status of being a police officer from the status of being married. Differentiation criteria may be regarded by the participants as a natural fact about people, or they may regard the criteria as dependent on other social statuses. 'Being a police officer' and 'being married' have social differentiation criteria. Race is different, for race is a status one is born with, not one that is acquired. Like the status of having royal blood, being an untouchable, or (in some societies) being a witch, having the right ancestors is an essential differentiation criterion for race. When thinking about differentiation criteria, it is important to remember that social roles are not all of a piece. Some are clearly defined and well-policed; others are more nebulous or contested. The differentiation criteria may even conflict with one another and create marginal or contested cases. Differentiation criteria for the social status of race, as it is treated in the United States, are a heterogeneous lot. As is often pointed out, the criteria have often been contested and have changed over time. In the nineteenth and early twentieth century, the 'one drop of blood rule' defined a person as non-white if they have any non-white ancestors. Whiteness is not so carefully monitored today. Phenotype is also an important differentiation criterion. A blue-eyed, blond-haired, light-skinned person with one Black great-grandparent is unlikely, today, to have the status of Black herself. Being Hispanic is a troublesome case for those who want to draw clear racial lines, because neither the ancestry nor the phenotypic markers are so clear. Cultural and behavioral markers, like speaking Spanish, are crucial for the differentiation of Hispanic status.

Behavioral expectations constitute a second way in which statuses are distinguished. They are the distinctive ways of acting created by uptake of the status and the norms that govern it. In the United States, we expect a police officer to carry a gun; we expect married persons to share income and residence. As with differentiation criteria, the consequences of uptake and the character of the norms vary enormously among social statuses. Some norms have the force of moral obligations and may be backed by the law (e.g. the obligation of married persons not to commit adultery); others may be little more than stereotypes. Uptake of some statuses may be central to how a person thinks of herself (e.g. being a mother), while others are contextual or temporary (e.g. being the referee of an intramural baseball game). The behavioral expectations of some statuses encompass authority and dominance hierarchies as well. Teacher/student, parent/child, manager/employee all carry expectations of dominance and authority. (Whether they *should* do so and the character such authority should take are important political questions.) When we turn to race, we find that the expectations for, and norms of, behavior are the most hotly contested aspects of racial status. Economic mobility and educational opportunities since reconstruction have wrought profound changes to the status of having a race in the United States. The Civil Rights Movement brought more, and it explicitly aimed at eliminating behavioral expectations having to do with the subordinate and oppressed status of being non-White. It remains an open question whether there are or should be any distinctive norms or expectations associated with being Black, Hispanic, Asian, White, or Native American. For our purposes here, it is enough to notice that insofar as discrimination based on racial distinction remains, these statuses still have a claim on us.

Considered as a social status, then, race does not differ significantly from ethnicity. Like race, one has to be born into an ethnicity; to be Danish one must have some Danish ancestors. In addition, ethnicities are distinguished by somatic features, culture, and language. Like race, the behavioral expectations of ethnicities arise from the identification with others of the same ancestors. Like race, both the differentiation criteria and the behavioral expectations of ethnicities are sometimes closely policed; in other times and places they are diffuse. What then, are we to make of Sundstrom's (correct) point that to conflate race and ethnicity is to fundamentally misunderstand American history and society? Two considerations modulate Sundstrom's point. First, since 'race' has no place in human biology, we cannot take the difference between the enslaved and the enslavers to be a natural difference. The difference can be nothing more than social status, and this already collapses 'race' into 'ethnicity.' The key player in American history is a particular status: being Black. We can admit that being Black and Danish in some small Midwestern towns are similar, *qua* social statuses, without thinking that being Danish had anywhere near the same significance for American history.

The second consideration is that we must bear in mind the difference between the conceptual commitments of those who make social distinctions and the conceptual commitments of the social scientists who study them. Social distinctions are often taken by the members of the society to reflect natural differences among people. Social scientists can recognize the social roles of such a society without agreeing that the social roles correspond to natural differences. The Azande, for example, believed that a witch had a unique substance in his or her abdomen. When a purported witch had been killed, the family could demand an autopsy to determine whether the person was really a witch. The Azande thus took 'being a witch' to be a natural fact about people, not a social status. E. E. Evans-Pritchard, who wrote about them, did not share this commitment. He did not confuse the Azande's concept of being a witch with his own. He treated 'being a witch' as a social status, and used it to explain cultural phenomena (Evans-Pritchard, 1937). Similarly, when we recognize the profound role that being Black has played in American history and society, or the differences in experience that attend this status, it is important to understand that Americans have taken race to be a natural fact about people. Historians and sociologists, however, need not and should not inherit this commitment.

To conceptualize races as social statuses distinguished by ancestry, somatic features, culture, or language is already to reduce race to ethnicity. This, again, is the sense in which the promiscuous realists have carried through the intra-theoretic reduction begun by nineteenth century opponents of racial theorizing. This reduction does not threaten to erase or minimize the social and historical importance of 'race.' There is an enormous difference between being Black and being Danish, both historically and in contemporary America. This difference is best understood as a difference in the historical or social significance of the status, that is, a difference in the behavioral expectations and consequences of being Black or Danish. This difference in content is recognizable precisely because both are in the same family of social statuses. To call this family of social statuses 'ethnicities' is simply to emphasize the break between the social concept of race and its nineteenth century predecessor.

6.4. RACE AND ETHNICITY IN MEDICINE

This essay began with the question of how 'race' is to be understood in medicine and the social sciences, given that there is no genetic or biological role for the concept. Can race be used as a proxy for genetic difference, and hence for genetic risk factors? This essay has argued that the nineteenth century concept of race was replaced by the social concept of race, and that a consequence of this intra-theoretic reduction is that 'race' and 'ethnicity' are not significantly different. Race is a kind of ethnicity with a particularly important history in the West. So, the original question becomes: Is ethnicity a relevant factor when assessing health risks? On this question, we can venture a qualified, affirmative answer for two reasons. First, ethnicity is a social status with differentiation criteria that depend on ancestry. Some ethnic groups have a relatively high degree of genetic homogeneity. This may be the result of a number of causes. The group may carefully police its boundaries or be so policed by others. Difference in language or customs may limit interbreeding. Accidents of migration history may produce geographic isolation and founder effects. In places where we can expect a strong correlation between ethnic status and genetic makeup, ethnic status will be useful as a proxy for genetic variation. Second, people who occupy the same social status are subject to the same norms. The norms that constitute differentiation criteria and the behavioral expectations may have health consequences. Discrimination in housing may force the Blacks of a particular town to live downwind of a smokestack belching toxins; Mexicans may be expected to spray pesticides on the lettuce fields. Given the right local conditions, there can be clinically significant correlations between ethnicity and health risks, even if the individuals are genetically heterogeneous.

Because ethnicity is a locally variable social status, the relevance of ethnicity to health can vary significantly among localities. Such variation may be either in the differential treatment or in the amount of genetic variation. The Danes in a small Minnesota town in may be genetically homogenous because they descend from a handful of homesteaders and have had little genetic exchange with outsiders. Such homogeneity is not true of everyone who might claim Danish status. It is almost certainly would not be true of the Danes in a nearby large city. The social significance of being Black is different in the suburbs of Atlanta, small towns in Louisiana, or the Chicago south side. Insofar as the differences in status result in different treatment, the health consequences of being Black will differ in these locations.

Such local variation in both genetic homogeneity and in differential treatment has important consequences for the epidemiological study of ethnicity and health. Many epidemiological studies use the census identification of race. For many variables of interest, the five or six standard racial categories are not homogenous with respect to either genetic variation or variations in treatment. A person may count herself Black (and be so treated by others) whether seven of her eight great-grandparents were European or all of her grandparents can be traced to the same village in West Africa. Moreover, the social status of being Black has different consequences in different locations. Generalizations about the health consequences of being Black, then, are meaningless unless we have reason to believe that there is relevant genetic or behavioral homogeneity among those who occupy this status. Moreover, since variation is local, statistically significant correlations between standardized races and health (or social) outcomes are likely to lack specificity. That is, the burden of the correlation will be carried by a sub-population that is relatively homogenous with respect to genotype or social treatment. To guard against a loss of specificity, researchers have to be sure that their ethnic categories identify groups that are locally meaningful. The downside is that such research is much more difficult; the upside is that it holds the promise of much better science.

There is no worry that genetic, health, or social research will resuscitate a nineteenth century conception of race. When the newspapers trumpet the discovery of a drug that works for Black Americans, this should not make us wonder whether biological races exist. That question cannot even be raised in the context of contemporary scientific theories. The two concepts that replaced 'race' – ethnicity and genetic variation – do all of the legitimate explanatory work done by 'race' and have none of its spurious consequences. Moreover, the concepts of ethnicity and genetic variation are sufficient to expose the real medical and social problems we face: problems of inequitable access to health care, unfair distribution of resources and opportunity, discrimination in housing and occupation, and differential genetic risk factors. Scientifically, if not in popular culture, race is a thing of the past.

Emory University, Atlanta, Georgia, USA

NOTES

[1] Some of these authors take the erotetic analysis of explanation to be both necessary and sufficient. The commitment of this essay is only that being a why-question is necessary, and that why-questions can be adequately analyzed in the manner of the erotetic model.

[2] The qualification is necessary because we have been operating under the assumption that explanatory autonomy is a necessary condition for ontological autonomy. It may not be sufficient, and if not, then promiscuous realism would not be entitled to its ontological commitment to the existence of races (at least at some times and places). This further issue is not important for the main conclusions of this essay.

REFERENCES

Boas F (1894/1974) Human faculty as determined by race. In: Stocking GW (ed.) The shaping of american anthropology 1883–1911: A Franz Boas reader. Basic Books, New York
—— (1911) The mind of primitive man. The MacMillan Co., New York
Cartwright S (1871) The diseases and physical peculiarities of the Negro race. New Orleans Med Surg J, 421–427
Collins JW Jr, David RJ, Symons R, Handler A, Wall SN, Dywer L (2000) Low-income African–American mothers' perception of exposure to racial discrimination and infant birth weight. Epidemiol 11:337–339
Dupre J (1993) The disorder of things. Harvard University Press, Cambridge, Massachusetts
Evans-Pritchard EE (1937) Witchcraft, oracles and magic among the Azande. Clarendon Press, Oxford
Fisher A, Wiebe V, Svante P, Przeworski M (2004) Evidence for a complex demographic history of chimpanzees. Mol Biol Evol 21(5):799–808
Folkes HM (1910) The negro as a health problem. J Am Med Assoc 15:1246–1247
Garfinkel A (1981) Forms of explanation. Yale University Press, New Haven
Harris S (1903) Tuberculosis in the Negro. J Am Med Assoc 41:834–838

Huxley J, Haddon AC (1936) We Europeans: a survey of 'Racial' problems. Harper & Brothers, New York

Jorde L, Wooding S (2004) Genetic variation, classification, and 'Race'. Nat Genet 36(Supplement) (11):S28–S33

Keita SOY, Kittles RA, Royal CDM, Bonney GE, Furbert-Harris P, Dunston GM, Rotimi CM (2004) Conceptualizing human variation. Nat Genet 36(Supplement) (11):S17–S20

Krieger N, Sidney S (1996) Racial discrimination and blood pressure: the cardia study of young black and white adults. Am J Public Health 86: 1370–1378

Linne CV (1758/1964) Systema Naturae. Stechert-Hafner Service Agency, New York

Lipton P (1991) Inference to the best explanation. Routledge, London

Livingstone FB (1962) On the non-existence of human races. Curr Anthropol 3(3):279–281

Massey D, Denton N (1993) American Apartheid. Harvard University Press, Cambridge Mass

McCauley RC (1984) Intertheoretic relations and the future of psychology. Philos Sci 53:179–199

Miller R (1987) Fact and method. Princeton University Press, Princeton

—— (2000) Half-naturalized human kinds. Philos Sci 67(Supplement): S640–S652

Patrinos A (2004) 'Race' and the human genome. Nat Genet 36(Supplement) (11):S1–S2

Risjord M (2000) Woodcutters and Witchcraft: rationality and interpretive change in the social sciences. SUNY Press, Albany, NY

Root M (2000) How we divide the world. Philos Sci 67(Supplement):S68–S639

Rotimi CN (2004) Are medical and nonmedical uses of large-scale genomic markers conflating genetics and 'Race'? Nat Genet 36(Supplement) (11): S43–S47

Sellars W (1963) Science, perception, and reality. Routledge and Kegan Paul, London

Sundstrom R (2002a) Race as a human kind. Philosophy and social criticism 28: 91–115

—— (2002b) Racial nominalism. J Soc Philos 33:193–210

Tishkoff SA, Kidd KK (2004) Implications of biogeography of human populations for 'Race' and medicine. Nat Genet 36(Supplement) (11):S21–S27

van Fraassen BC (1980) The scientific image. Oxford University Press, Oxford

Wade N (2004) Articles highlight different views on genetic basis of race. The New York Times, Wednesday, 27 October 2004, p A18

KELLY C. SMITH

7. TOWARDS AN ADEQUATE ACCOUNT
OF GENETIC DISEASE

'For every complex problem there is a simple, easy to understand, incorrect answer' —Albert Szent-Györgyi

7.1. INTRODUCTION

It is common practice now to describe biological traits, especially diseases, as 'genetic'. Surprisingly, however, there has been little careful study of precisely what such descriptions actually *mean*.[1] In fact, the different meanings of the term 'genetic trait' are nearly as numerous as the occurrences of the phrase itself. Moreover, most of these accounts are so uncritical that they fail to see even very obvious shortcomings.

To be sure, a large part of the problem stems from the fact that the etiology of biological traits is typically extremely complex and, like all complex causal processes, involves a very large number of causally relevant factors which interact with each other in intricate ways. Indeed, biological processes add an additional level of complication which makes causal analysis of organic traits particularly frustrating: an extremely high degree of *variability* in the causal factors involved.[2] Levels of biological variability are so high that it is something of a truism to claim that each and every biological organism is importantly unique. This 'mega-complexity' is such a prominent feature of living things that it's often taken to be what differentiates biology from other sciences:

complexity in and of itself is not a fundamental difference between organic and inorganic systems. The world's weather system or any galaxy is also a complex system. On the average, however, organic systems are more complex by several orders of magnitude. (Mayr, 1988, p. 14)

When attempting to explain the dynamics of any complex causal system, two separate problems present themselves. The problem of *causal connection* is the problem of how to determine 1) *which* factors are causally involved in the outcome and 2) the character of their causal interactions. The problem of *causal selection*, on the other hand, is the problem of how to select the single causal factor most suited to the causal explanation of the outcome from amongst those which are causally connected.[3] In other words, causal connection involves questions about how the causal matrix actually looks, while causal selection concerns questions about how we can *explain* that matrix.

83

H. Kincaid and J. McKitrick (eds.), Establishing Medical Reality, 83–110.
© 2007 *Springer.*

The problem of causal selection was first carefully analyzed by J.S. Mill, who concluded:

Nothing can better show the absence of any scientific ground for the distinction between the cause of a phenomenon and its conditions, than the capricious manner in which we select among these conditions that which we choose to denominate the cause. (1859, p. 214)

The temptation to adopt a defeatist attitude with respect to such complex causal processes is a strong one. Yet there is also a legitimate practical purpose for such 'explanatory shortcuts' – if nothing else, we can't describe the causal situation exhaustively every time we wish to claim something about the cause(s) of disease. Moreover, these shortcuts are relatively harmless as long as we keep in mind precisely what information they do and do not convey about the causal system in question. Problems can arise quite easily, however, as when a vague or overly simplistic criterion is used for causal selection or when the criterion used is misinterpreted by one's intended audience. It is crucial, therefore, that causal selections always be done on the basis of an explicit criterion which has been critically evaluated. If we identify a disease as genetic, we should be able to give a clear account of precisely what this entails and why such a description gains more through concise transmission of important causal information than it loses through incompleteness.

Unfortunately, this is almost never the case in current descriptions of genetic traits. Typically, there is no explicit criterion at all. When criteria are indicated or can be inferred, they are often so simplistic that the explanations they produce distort more than they inform. This is a curious state of affairs, to be sure. After all, claims about the genetic status of traits are routinely made by some of the most gifted scientists in the world. How could they be so careless?

Scientific researchers are trained to be exceedingly careful about the data they present and the conclusions they draw from that data. They focus almost exclusively on causal connection questions and don't devote much time to considering causal the selection problem. Thus, when they do engage in causal selection, they give free rein to a relatively uncritical kind of gene-centric metaphor.[4] Essentially, gene-centrism is the view that the genes[5] are the most important explanatory factors in biology because of their unique causal powers – whether in controlling individual ontogeny (development), accounting for abnormal functioning in adults (disease), or explaining changes in populations over time (evolution).[6] This tacit metaphor about the role of genes in biological organisms goes a long way toward explaining the uncritical and enthusiastic 'genes for' locutions which are so common in both the lay and professional press (Allen, 1980).

It will be difficult to develop an adequate account of a genetic trait as long as gene centrism looms in the background. Thus, our first task must be to show exactly why gene-centrism is unsatisfactory. I will start by analyzing a few of the arguments which are thought to support gene-centrism. These are all shown to depend on such highly oversimplified pictures of biological processes that they are ultimately indefensible. Hopefully, this critique will serve to undercut some of the intuitive appeal of genetic explanation of traits in general, allowing us to precede from that point free from the shadow of simplistic gene-centrism. Next, I will set out a series of traditional philosophical attempts to solve the causal selection problem. It will be shown that each account is inadequate as a general criterion of causal selection for biological traits. Some of these accounts, especially manipulability, might have limited utility in biological explanations. However, this could only occur through the addition of a thoroughly statistical analysis of variation within populations. Finally, I will offer just such a statistical approach, the 'epidemiological' account of genetic traits. The epidemiological account employs statistical analysis of variance to clarify what a 'genetic trait' should and should not mean in complex biological systems. Approaching causal selection statistically avoids many of the criticisms leveled against the alternatives, though arguably at some cost to our intuitions about causal explanation.

7.2. GENE-CENTRISM EXAMINED

7.2.1. Gene-centric Arguments

I begin in what may initially seem a roundabout fashion by discussing three common 'arguments' for gene-centrism. Although these are not criteria for the causal selection of genes per se, this is nevertheless a necessary first step since this type of thinking tends to deflect any call for the development of explicit selection criteria or critical scrutiny of existing criteria. As it happens, these arguments all rest on inappropriate, even gross, oversimplifications of the relevant causal processes – a point which might be recognized more readily but for the fact that they are rarely stated as baldly as they are below.

7.2.1.1. Genes came first   The argument goes like this: Since the earliest entities capable of the uniquely biological process of *replication* were nucleic acids, all other entities in biology (organisms, environmental factors, etc.) are logically secondary to the genes.

Of course, since the origins of life are obscured by 4 billion years or so of intervening history, we will never know for certain what the earliest evolutionary steps actually were. On the other hand, our current understanding of chemical evolution makes the claim that genes came first debatable. It's certainly true that, in modern systems, genes are an indispensable part of cellular chemistry, as well as paradigm replicators. While it's intuitively appealing to think that this has always been the case, many biologists working on the origins of life think that the earliest replicators were relatively simple sets of autocatalytic chemicals lacking nucleic acids.

For example, sets of protein-like amino acid polymers (proteinoids) can be synthesized relatively easily in primordial conditions (Fox and Dose, 1972). When these become encapsulated, the result is an entity with many of the properties we associate with life. Such 'protocells' can maintain homeostasis, ingest and metabolize other chemical compounds, and replicate themselves in a way that passes on their unique traits. Some might even have been capable of crude photosynthesis via ingested pigments (Oparin, 1924). These protocells would certainly be subject to evolution by natural selection and may even exhibit an intrinsic capacity for rapid increases in complexity that would make the evolution of life much more mundane and predictable than is generally thought (Kauffman, 1993). On this view, genes came late to the party – and when they did come, their role was simply to increase the efficiency with which heritable information was already being passed on. Therefore, it may well have been non-genetic causal factors (epigenes) rather than the genes that came first.

Presumably, the appeal of the 'genes came first' argument derives from the fact that genes were originally *independent* of extra-genetic causal influence. It seems likely, however, that neither genes nor epigenes[7] came first in the sense that the earliest true lifeforms were integrated systems composed of both genetic and epigenetic components.[8] However, even if genes did 'come first', this fact does nothing to refute the current intimate interdependence of genes and epigenes. Thus, whatever the facts are concerning the very earliest stages of evolution, it seems unlikely they can be used to generate a plausible justification of gene-centrism in a modern context.

7.2.1.2. Only genes are transmitted   The gene-centric argument here goes like this: Since only genes are transmitted from one generation to the next, genes are the only candidates for explaining how traits are realized in a particular generation (Williams, 1966; Dawkins, 1976).

The picture this argument paints is one of a tenuous connection between successive generations via a 'genetic bridge' of naked DNA. But nothing could be further from the truth. For one thing, many organisms are asexual and never pass through the bottleneck of a single celled gamete stage. It's hard to defend the claim that only genes are being transmitted by the growth of a complex, multicellular runner, as is common in plants. The case seems better in sexual organisms, but only because we often adopt a ridiculously simplistic picture of the gametes. The sperm bears the brunt of this caricature, often being derided as 'nothing more than a bag of DNA with a tail'. However, sperm do contain cytoplasm with heritable epigenetic structures. For example, in many organisms sperm are known to transmit the centriole, a protein structure indispensable for proper genetic segregation. Certainly when one considers the egg, the argument loses all plausibility. The egg is hardly the 'homogeneous mass of goo with genes' it is sometimes assumed to be – rather, it's an incredibly complex array of a vast number of chemical compounds arranged in intricate patterns. This array has its own dynamic potentials which can't be reduced to genetic control – as evidenced by the fact that fertilized eggs can frequently sustain several perfectly normal cell divisions even when the nucleus (with its genes) is completely removed (Gilbert, 1997). In short, there are any number of epigenetic causal factors transmitted along with the genes whose causal influence is indispensable, thus it's simply false to claim that genes are the only heritable causal factors.

7.2.1.3. Genes are uniquely accurate replicators   The argument goes like this: In order for evolution to occur, there must be *heritability*. That is, there must be something which can be said to replicate and thus survive multiple generations – without this, the effects of natural selection could not accumulate over time. Genes replicate themselves so accurately that they are, in effect, the only real candidates for an evolutionary vehicle of information.

Some of the shortcomings of this view are fairly technical[9] but three basic problems can be set out fairly simply. For one, if the genes are uniquely accurate, this is due largely to the causal role played by *epigenetic* factors. In particular, the high copy fidelity of DNA is accomplished through epigenetic mechanisms which correct replication errors. Secondly, there are at least some epigenetic structures (e.g., microtubule organizing centers such as centrioles) which seem to replicate quite accurately – it is an open empirical issue whether they are more or less accurate than genes. Lastly, the threshold of accuracy below which natural selection can have no effect is very low indeed. This means that *any* heritable variation with advantageous effects will tend to be preserved, even if it has a low (absolute) heritability. Of course, all else being equal, the more accurately a variant replicates, the better it will be preserved. Yet even on the assumption that genes always replicate more accurately than epigenes, the most that can be said is that evolution will *tend* to favor genetic systems over epigenetic ones when parallel transmission systems for the same variation exist in competition with one another. That is, natural selection will tend to eliminate systems with low *relative*, not absolute, heritability. Replicators with low absolute heritability are neither impossible nor unlikely – in fact; nature is replete with examples of highly inefficient processes which nevertheless persist through evolutionary time simply because no better alternative has presented itself.

7.2.2. A Gene-centric Response?

In order to avoid the charge of attacking a straw biologist, I want to examine directly the kind of response I often hear from gene-centrists at this point. Of course, they say, lots of things besides the genes are going on inside the cell – but no biologist denies this. On the other hand, everything *starts* with the genes and proceeds from there. Since the genes *construct* the enzymes and cellular machinery involved in their own processing, it is misleading to represent these factors as somehow *independent* of the genes. Genes make everything else in the cell, though it's certainly true that they also depend on these products in many complex ways for their proper functioning. The crucial point, however, is that whatever other factors may be involved, genes *drive* the whole process.[10]

But consider the preceding discussion. Genes very likely did not evolve first – they were either preceded by epigenetic replicators or were integral parts, along with genes, of the first lifeforms. Genes are not transmitted independently from one generation to another and thus it's simply not true to say that upon them alone falls the job of building up a new organism de novo. Genes may or may not be uniquely accurate replicators, but even if they are, this does nothing to preclude the very real possibility that epigenetic factors can exhibit similar properties. Why should we say that genes drive the epigenetic processes rather than vice versa? In short, the intuitive case for the priority of genes in general cannot withstand the light of careful scrutiny.

7.3. TRADITIONAL CAUSAL SELECTION TECHNIQUES

Having hopefully at least dulled the intuitive appeal of gene-centrism, I will now move on to discuss several classical techniques which either have been or might be used to identify biological traits as genetic. I divide the techniques into three groups. The first group, which I call 'intuitive techniques', have actually been used (often implicitly) to try to make sense of genetic causation. These fail completely as defensible analyses of genetic traits. The second group of techniques was developed mostly as approaches to casual explanation generally and often have never been explicitly applied to genetic causation contexts. These approaches are more defensible in general and may even have limited utility in the analysis of genetic traits (though only with extensive revision). Finally, I consider two techniques from medicine which are often cited as helpful in discussions of disease causation but which also fall far short of what is required.

7.3.1. Intuitive Techniques

7.3.1.1. Genes are causally involved in traits   Here, a genetic trait is thought of as one in which genes play a significant causal role.

This seems to be the sort of implicit thinking behind many of the claims, so common in the press, concerning the discovery that this or that disease is 'genetic'. But this account is far too blunt a tool to support such claims. For one thing, it provides no justification for why the causal involvement of genes serves to select them as *especially* explanatory while the causal involvement of various epigenes does not. At the very least, any condition influenced by genes will also be influenced by proteins, since it is only through the construction of proteins that genes exert their effects (see Smith, 1992).

Furthermore, the involvement of genes in biological traits is so pervasive that it seems difficult to envision a trait in which genes are *not* involved. On this account then, every trait will be labeled as genetic – a position actually endorsed by some medical researchers with respect to disease. Yet it is hard to see how a description of this sort conveys any useful information, precisely because the trait in question is universal.[11] All it does is establish that genes are essential factors in biological processes – a fact with which no one would argue. It does not establish that genes are any *more* important than other causal factors, though this seems implied by the 'genetic' label. Truth be told, we have exactly the same grounds, on the basis of causal involvement alone, to describe all biological traits as 'protein-based' or 'organic'. These descriptions seem silly precisely because they do not convey any new information. So, if genes are to be causally selected as the best explanation for some trait, this must be based on unique characteristics of their causal involvement, not merely the *fact* of their involvement.

7.3.1.2. Genes produce traits 'directly'   Here, a genetic trait is thought of as one in which the causal connection between the genes and the trait is (relatively) direct.

Although this is the only intuitive method of causal selection which has actually been explicitly discussed in the context of genetic traits (Hull, 1981; Gifford, 1990), it is also one of the more dubious. To begin with, the concept of 'directness' seems inherently fuzzy. A good working definition might be that a causal process is direct to the extent that it involves (relatively) few steps between initiation (by the genes) and final outcome (the trait). However, whether this applies to a trait seems entirely a matter of how one chooses to describe the causal process. The account typical of an introductory biology textbook says that genes make an mRNA transcript, which is translated into protein, which then has an effect. Most people would probably allow that this seems fairly direct, though this seems an inherently subjective judgment – how many steps can one allow before labeling a process 'indirect'?

Consider a more detailed account of the process: complex regulatory machinery activates the genes in response to subtle epigenetic signals, a multitude of enzymes unpack and transcribe the DNA, the RNA transcript is spliced and processed as appropriate, the spliced transcript is transported to the ribosome and translated by the machinery there, the emerging protein is folded into one of several confirmations, and finally the protein is combined with other carefully constructed proteins, all of which are transported to the region required. This is a process which may involve hundreds of enzymes and dozens of complex environmental feedback systems. It may span the entire length of the cell and utilize a significant portion of the cellular machinery. And this is just a description of what happens in a single cell – we haven't bothered to say anything about cell-cell interactions, much less anything at the level of tissues or organs. It seems much less plausible to describe the process now as direct. However, note that all that has changed is the level of detail in our description. Directness accounts then, are entirely dependent on a highly simplified portrayal of trait etiology.

Another problem with directness is that, however direct one thinks genetic processes are, choosing to describe a particular trait as genetic is a comparative judgment – one is saying that the genes are *more* directly linked to the outcome than alternative explanations citing epigenes. It is often simply asserted that genes are the most direct – as when Hull claims that 'genes pass on their structure in about as direct a fashion as can occur in the material world' (1981, p. 151). However, since the directness with which *epigenetic* factors pass on their structure is actually an open empirical question, we have no good grounds for the claim that genes are more direct.

Lastly, there is the truly fatal fact that directness is just not relevant to causal selection. Why should we care how direct the process is which links a cause to its effect? It seems directness advocates implicitly assume that, the more direct a process, the more accurate the production of the end product (a trait). Again, this has a certain intuitive appeal. However, the idea that the directness of a process is an indicator of its accuracy is highly dubious in light of the fact that genetic replication is so accurate precisely because it is so *indirect* – in particular, because it involves elaborate (epigenetic) error correction mechanisms. Indeed, nature seems fond of positively Rube Golbergian processes, anything but direct, which nevertheless faithfully reproduce some trait of importance.[12]

7.3.1.3. Genes are necessary for traits   Here, a genetic trait is thought of as one for which genes are necessary (i.e., without the gene, the trait could not occur). Paradigmatically, when knockout studies show that inactivation of a gene (by mutagenesis, etc.) results in alteration of a trait, we have grounds for saying that the gene seem necessary for the trait and thus that it's genetic, at least in some sense.

But does a knockout experiment really establish the necessity of the genes? One thing we must always keep in mind is that there may well be cases of the trait occurring in the absence of the genes – there are often multiple causal routes to a particular biological outcome.[13] Just because another pathway to the trait was not used (for whatever reason) in a particular knockout study does not imply that there are no other pathways. Of course, we do not want to take this point too far as it is never possible to prove the nonexistence of something (in this case, an alternate causal path). However, negative proof difficulties aside, there are still reasons to question the strength of the evidence for necessity. For one thing, we often forget how little information we actually have about variance in causal factors at the population level (see discussion of cystic fibrosis). Without such information, assessing the proportion of cases where the trait is *not* accompanied by the genes is often mere guesswork. Moreover, most genetic studies are performed on highly inbred lines of organisms, which has the effect of artificially suppressing the variation in causal factors (both genetic and epigenetic). While this makes the effects of genetic manipulations much easier to detect, it also significantly lowers our confidence that the genes will have the same effect in natural, more variable, populations. Finally, when we insist, as we often do in modern medicine, on equating a trait with its known genetic basis, it is hardly surprising that our studies then 'confirm' that the trait is genetic (Wulff, 1984).

The necessity technique also fails, as did the causal involvement account, to differentiate the influence of necessary genes from other necessary factors. Even when we have very good grounds for believing that genes are truly necessary for a trait, it will always be the case that many other factors besides genes share this distinction. We might wish, as Gifford (1990) suggests, to refine the necessity criterion by excluding all factors which are universal (or nearly so) from consideration. By 'universal', Gifford means a factor which is present in all cases of both the occurrence and non-occurrence of a trait. To use his example, the absence of 1000 degree temperatures might well be a necessary condition for having a given trait. However, since it is also a requirement for life as we know it, and thus of any trait whatsoever, it does not constitute a legitimate alternative to the genes as an explanation of a particular trait. This narrows the field of alternative necessary factors a bit, but not much. The reason for this is that there will usually be a great many necessary conditions for a trait which are not, properly construed, universal. For example, while the *existence* of a protein within a cell might well be universal, its precise concentration and distribution could be the salient causal features with respect to a given trait – and these are hardly universal (Smith, 1992).

So, we have seen that none of the intuitive accounts of causal selection can plausibly defend the concept of a genetic trait in general. The causal involvement and necessity accounts fail mainly because they do not provide grounds to differentiate the influence of genes and epigenes. The directness account has many difficulties, chief of which is the fact that it's not even a relevant consideration for causal selection.

7.3.2. Philosophical Techniques

The next techniques of causal selection have been developed fairly extensively in the philosophical literature on explanation. However, none has been carefully applied to the special case of causal selection in biology. Sadly, it's quite common for beautiful theories in philosophy of science to come to grief when they encounter biological complexity and these techniques are no exception to that tradition.

7.3.2.1. Genes are 'abnormal' factors for traits   Here, a genetic trait is thought of as one where genetic factors can be picked out as the 'abnormal' ones.

This is a technique originally explored in the analysis of legal responsibility – in particular, in contexts where we need to assign *blame* (Hart and Honore, 1959; Hilton, 1988). In such contexts, we have (at least arguably) a moral concept of 'normality' that we can apply independent of any consideration of variability. This is important, because without such norms, statistical analysis of normality will produce odd results. For example, it might be that almost everyone in dire financial need cheats on his taxes. Such behavior would thus be 'normal' in the purely statistical sense, though a court would be unlikely to exonerate a defendant who admitted such cheating yet argued that it was the financial difficulties, not his actions, which constituted the abnormal causal factor in the crime. Therefore, although it will often be the relatively rare causal factor which is selected, it will not be selected because of its rarity.

The real problem, however, is that we have nothing like an independent theory of normality in biology as we could be said to have in moral contexts.[14] In fact, it's not even clear how we should go about creating such an account in a discipline where *variance is the norm.* Finally, even if we eventually develop an account which clears these hurdles, its primary applicability would be to more general concepts like 'health' and 'disease,' where there is an intuitive appeal to something like a 'norm' (e.g., proper functioning, etc.) In short, the concept of normality doesn't work well when it comes to the analysis of specific biological traits, unless by 'norm' we are ultimately just appealing to statistical regularities (in which case we need a theory which does this explicitly).

7.3.2.2. Genes are the more 'manipulable' factors in traits   Here, a genetic trait
is thought of as one where the genes are manipulable in such a way as to produce,
prevent or alter the trait.

The classic version of this approach in philosophy is Collingwood (1938), though
it has received some attention in the particular context of genetic traits as well
(Magnus, 1996). It has great intuitive appeal and seems an eminently practical
approach. In particular, many find it a compelling account of diseases, since these
are traits we are generally much more interested in preventing than in explaining
purely for the sake of scientific understanding.

However, one underappreciated yet very pernicious problem with the manipulability account is that it equivocates between two different kinds of manipulability –
manipulability *in practice* and manipulability *in principle*. Arguing that we should
explain a trait by the factors which we can manipulate *in practice* – those we can
actually tinker with using present or soon to be developed technology – makes the
most forceful appeal to our practical sensibilities. Yet, as of this writing, almost
no diseases would truly be classified as genetic in this sense. As a technological
optimist, I tend to view this as a temporary state of affairs, but unless we are
willing to withhold attributions of 'genetic' until we have the actual technology to
manipulate diseases via the genes, this is a real problem.

Using manipulability in practice also makes the nature of a trait relative to
our present knowledge in counterintuitive ways. For example, a disease may be
environmental now, genetic later on when we learn the appropriate genetic intervention, and then epigenetic later still when a more effective non-genetic treatment is
discovered (see Stern, 1973 and Burian, 1981–82). Of course, knowledge-relativity
is a feature of many explanation theories and isn't necessarily a fatal flaw. What
makes it damning for manipulability in practice is that the knowledge on which the
explanation hinges need not be anything about the causal process in question. For
example, suppose we knew everything there is to know about the causal processes
involved in colon cancer: the genes involved, their effects and interactions, all the
environmental effects and their variance, etc. With such perfect causal knowledge,
we could probably devise a genetic intervention which would prevent the disease
with certainty. Yet we must also be able to *implement* this intervention in order
for the trait to be considered genetic. We could easily be thwarted in our efforts
because we lack knowledge only tangentially related to colon cancer – for example,
the technology to precisely vector our genetic changes to the target cells in the
precancerous polyps. It seems not just counterintuitive, but contrary to the purposes
of causal explanation, to argue that such a trait isn't genetic simply because we
lack a distantly related technology.

The obvious response a proponent of manipulability can give to this sort of objection
is to loosen the requirement that traits be manipulable in practice. If we say that a genetic
trait is one which is alterable *in principle* via genetic means, then we do not have to
deal with knowledge relativity. However, the further we go down this road, the more
we give up the *practical* appeal which motivated the manipulability approach in the

first place. After all, who really cares whether a disease can be manipulated genetically if we will not know how to do this for 300 years?

A more important problem is that this account casts a very wide net indeed. After all, given the intimate role played by the genes in all biological systems, *any* trait will be genetically manipulable in principle. But then what does it profit us to describe all traits as genetic, regardless of the particulars of their etiology? To make matters worse, any trait that is genetically manipulable in principle will also be *epigenetically* manipulable in principle. If nothing else, for any trait in which manipulation of a gene has an effect, we can always simply manipulate the gene *product* to produce an identical effect. Therefore, while manipulability in practice contaminates our explanations with a very unwelcome form of knowledge-relativity, manipulability in principle is so broad as to be useless from a practical standpoint.

7.3.2.3. Genes are the 'precipitating factors' for traits   Here, a genetic trait is thought of as one where the factor which varies immediately prior to the effect, thus 'precipitating' it, is a genetic one.

This technique has a lot of appeal when applied in some contexts (Ducasse, 1924; Ryle, 1949).[15] To take the classic example, suppose we wish to analyze the cause of a barn fire. We find that the fire occurred immediately after the application of a lit match to some hay in the barn. Of course, a great many factors were causally involved, even necessary, to bring about the fire: the presence of oxygen and combustible materials, the absence of water, the laws of oxidation, etc. Nevertheless, we would identify the lit match as the *most* explanatory cause of the fire, though we would certainly allow that other factors were causally important in various ways. The reason for this intuition, according to the precipitating factor technique, is because the lit match varies in a way that *immediately precedes* the occurrence of the fire (as, for example, the presence of oxygen does not).

Yet how, precisely, is one to construe the phrase 'immediately precedes?' Philosophers point out that there are a great many causal steps intermediate between the lit match touching the hay and the burning barn.[16] Thus, one could argue that it is not really the lit match meeting hay which immediately precedes the fire, but rather the air currents which serve to supply oxygen during subsequent developments. After all, these air currents are not only closer in the causal sequence to the effect, but are also necessary – if the match had been applied to some hay under an airtight bell jar, the barn would not have caught fire. Of course, this strikes those of us with a practical bent as missing the point since it seems obvious from the outset that it's the match which is really the object of interest.

Precisely why is this? We intuitively reject the air currents in favor of the match-hay interaction because we assume that oxygen-rich air currents are essentially universal within barns (as are the chemical laws governing oxidation, etc.). Our intuitions would certainly seem much less secure if most barns contained lit matches, but only a very few of them contained oxygen.[17] If we wish to be careful, therefore, the precipitating factor account must be relativized to some *population* (in this case, barns with a glut of oxygen and a dearth of lit matches).

So, could we use a population-relative precipitating factor account to defend the genetic status of traits? Perhaps in some cases we can, but we must be very careful. Barns are ridiculously uncomplicated things compared to biological organisms. Even within specified biological populations, we rarely have a very detailed picture of the causal factors at play in the genesis of a trait; much less can we adequately characterize their variability. Nor can we mistake our causal ignorance for evidence that unknown factors do not exist or are not variable – given the high levels of variability typical of biological systems, the burden of proof should be placed squarely on those claiming *in*variance. Besides, there is frequently good evidence of variability in unknown epigenetic factors. For example, women with the BRCA1 gene are at higher risk for breast and ovarian cancer, but it is far from certain that any particular woman with the gene will develop cancer. Similarly, two people with the same mutation in the CFTR gene may experience entirely different symptoms – one developing a fatal case of cystic fibrosis, the other only chronic bronchitis. In both cases, this is presumably because some unknown epigenetic causal factor intervenes.

The precipitating factor account thus makes sense only when it is relativized to some particular population in which we have knowledge about the range of variability in causal factors. Because the philosophical analysis has tended to focus on cases (burning barns) where variability of most causal factors is low or nonexistent, this requirement has not been explicitly recognized. Yet clearly the precipitating factor account requires, at the very least, some criterion for assessing variability (more precisely, for assessing the *covariance* of the causal factor with its putative effect). It will also have to confront the fact that its explanations of biological traits will be relative both to the population chosen and the level of our causal knowledge.

7.3.2.4. The 'sufficiency' account   Here, a genetic trait is thought of as one for which genes are sufficient (i.e., the presence of the gene guarantees the development of the trait).

Strictly speaking, there is no such thing as a gene which is sufficient for a trait. Even if we are talking about very simple traits (say, the possession of a particular protein), we have seen that the genes cannot achieve this effect without a great deal of help from epigenetic factors. Of course, this is not news to biologists and thus it would be in keeping with the principle of charity to assume that they have something a bit different in mind.

What biologists probably mean when they make this kind of argument is something like the following: All else being equal (ceteris paribus), the existence of the genes will bring about the trait in question. In other words, the influence of the genes becomes apparent once we establish a uniform epigenetic background in which to manipulate them. However, the ceteris paribus clause can cut both ways: for every explanation of a trait as genetic, it's possible to create a complementary explanation of the same trait as epigenetic (and on precisely the same evidentiary basis). The difference between the two accounts is simply whether one chooses to have the ceteris paribus clause restrict epigenetic or genetic variance (Smith, 1992).

If we want to use this kind of account to defend the description of a trait as genetic (contra epigenetic), then there must be some independent grounds for the claim that the ceteris paribus clause should be taken to prevent epigenetic rather than genetic variance. One possibility would be if the epigenetic conditions were universal and invariant – then the imposition of the epigenetic restrictions would only rule out counterfactual examples which are not of immediate practical significance. Here we must, as previously discussed with the necessity account, take great care in how we characterize the causal factors (and thus their variance) to make sure that we are capturing the causally salient aspects. Unfortunately, there are not many biological traits where all the causally important epigenes are universal.

The second possibility is that the epigenetic factors are invariant, not universally, but rather within some population of interest. This seems to be the kind of account that is required. However, this does not establish genetic sufficiency in the strict sense – even within the population in question. This is because sufficient causes in biology will typically be complex *combinations* of causal factors. In any event, the classic discussions of sufficiency give little clue as to precisely how we are to go about constructing a population-relative, variance-sensitive account of causal selection

So what does the sufficiency account really accomplish? First, it fails to establish strict sufficiency. It does show, given the fact that genetic intervention affects the trait, that genes are causally involved. But then, we've already established that this is not adequate grounds for causal selection. At best, it can identify when genes are 'sorta sufficient' – provided we somehow factor in a limited variance in causal factors within a population. It would appear that the sufficiency account needs help from another source to realize its potential.

7.3.3. Medical Techniques

7.3.3.1. Koch's postulates   While he was not concerned with the question of genetic disease, epidemiological pioneer Robert Koch did develop three key postulates to help us select the explanatory causes of disease. Koch argued that we are justified in saying a particular factor (pathogen) causes a particular disease whenever three basic conditions are met:

1) The pathogen is always found in individuals with the disease.
2) The pathogen is never found in individuals with conditions other than the disease.
3) The pathogen always produces the disease when introduced into healthy individuals.

Note that postulate #1 is simply a requirement that the pathogen is necessary, while postulates #2 and #3 are requirements that it also be sufficient.[18] It is certainly true

that these postulates served Koch and others well in the early field work for which
they became famous. The pathogenic diseases Koch studied are probably uniquely
well suited to this type of analysis, involving as they do a clearly identifiable
infectious agent with well-defined and dramatic onset of a stereotypical set of
symptoms. But even for infectious diseases, these postulates are, strictly speaking,
too strong. For example, many people are infected with the TB bacillus yet never
exhibit the disease, but we do not conclude that TB is not caused by the bacillus.[19]
In any event, infectious diseases are not representative of disease in general and
certainly not of genetic disease.

7.3.3.2. Hill's epidemiological criteria   Another traditional approach in medicine
uses something like the epidemiological criteria developed by Sir Austin Hill. This
is essentially a more sophisticated version of Koch's postulates, relying on eight
different criteria:
1) *Strength*
2) *Consistency*
3) *Specificity*
4) *Temporality*
5) *Biological gradient*
6) *Plausibility*
7) *Coherence*
8) *Analogy*
This is undeniably an improvement over Koch's original formulation, since it allows
for a much more nuanced description of the causal relationship between a particular
factor and the disease. However, where Koch's postulates will usually at least
yield a clear answer, Hill's criteria will often not yield much of an answer at all.
In particular, Hill provides no clear method of *ranking* or *weighting* the various
factors. Thus, it is unclear what we are to make of a putative causal agent which
scores well on one criterion but poorly on another, as will be common in putative
genetic diseases where many different factors interact to influence the likelihood
and severity of the condition.

7.4. THE EPIDEMIOLOGICAL ACCOUNT

7.4.1. Intuitions and Questions

We are now in a position to develop my own analysis of genetic traits, which I call
the epidemiological account. In doing so, I will focus on genetic disease rather than
genetic traits more generally. One reason for this is that human diseases typically
have a complex etiology which highlights the importance of the kind of statistical
analysis I employ. Another reason is that, from a purely practical point of view, we
are most intimately interested in disease since we are all subject to it. In any event,
a truly complete account of disease is far beyond the scope of this work. Rather,
I am assuming the more modest goal of outlining an account of disease causation
which is *minimally adequate*. Descriptions of a trait within such an account must 1)

be consistent with the sort of complex causal networks with high variability we routinely encounter in biology, 2) impart sufficient information to be of practical utility and 3) distort the complete causal picture as little as possible.

I call my account the epidemiological account because it draws its inspiration from the field of epidemiology. Like epidemiology, it is an analysis of disease that crucially depends on statistical methods applied to populations rather than individuals. It is unlike epidemiology, however, in one crucial particular: while epidemiologists invest a great deal of their effort in finding causes to associate with disease (solving the causal connection problem), my concern is strictly with the *explanation* of disease (solving the causal selection problem). For the purposes of these discussions, I simply assume that we have already identified all the significant causal factors. To be sure, I do not want to appear to underestimate the complexity of this task, but problems of causal connection are very different from those of causal selection and have already been the focus of extensive analysis.

It is important to appreciate that epidemiological analysis is fundamentally an examination of the properties of *populations*.[20] There are two important implications of this that we must be careful to keep in mind. First, it is entirely possible that epidemiological analysis of one population will label a disease 'genetic' while a similar analysis of another population will label the same disease 'environmental' (see e.g., Burian, 1981–82). From the point of view of epidemiological analysis, this is because the two populations actually represent two importantly different phenomena, since what is actually being analyzed is the variance within populations, not a disease known independently of context.

Secondly, because the analysis is ultimately of the variance within populations, the resulting explanation is also about populations, not individuals. We might be able to say that a given type of cancer in a given population is genetic, but will not be in a position to say whether any particular case of cancer is genetic. I freely grant that this is counterintuitive and requires a major shift in how we think about disease. However, I also argue that such population-relativity ultimately cannot be avoided in complex causal analysis of the sort we have in mind here, *whatever* method we adopt (see discussion of relativity below). The virtue of the epidemiological account is that it makes this aspect of complex causal analysis explicit.

My account is designed to satisfy the two most basic and widely held intuitions about disease causation. Simply put, these are:

1) *The individual intuition*: If a disease is genetic, this must mean that those with the gene are more likely than not to develop the disease.
2) *The population intuition*: If a disease is genetic, this must mean that cases of disease more likely than not causally involve the gene in a significant way.

However, before we are in a position to pursue these intuitions in more detail, we will first have to develop some statistical machinery by examining a series of cases.

7.4.2. Developing the Account through Cases

Before attempting to apply the basic intuitions about causal explanation above, I first need to develop some epidemiological concepts. When we explain a disease as 'genetic', we wish to convey that, in this particular case, the gene is somehow the predominant explanatory factor. However, it is crucial that the tests we employ in establishing this do not themselves subtly beg the question by assuming what sorts of causes predominate in explanation. We must, for example, be very careful with language, speaking only of 'causal involvement' and not 'the cause' – while the former simply asserts a causal connection, the latter implies some kind of causal selection.[21] The basic method we must use is thus one which examines the correlation between, on the one hand, the presence of a gene we know to be causally connected to the disease and, on the other hand, a disease state.

However, this very general correlation question will be answered in different ways and with different implications depending on the population to which it is posed:

1. What percentage of the entire (global) population will develop the disease in a way that casually involves the gene? This I call the *global question*.
2. What percentage of the subpopulation consisting of those with both the disease and the gene developed the disease in a way that casually involved the gene? This I call the *diagnostic question*.
3. What percentage of the subpopulation having the disease developed the disease in a way that casually involved the gene? This I call the *testing question*.
4. What percentage of the subpopulation having the gene will develop the disease in a way that casually involves the gene? This I call the *prognosis question*.

We will need to look at particular cases to develop the tools we will need to answer these questions and, later, apply our intuitions to the causal selection issue. So suppose we look carefully at a population of 10,000 people where 12.5% (1250) have a particular disease and 10% (1000) have a gene thought to be causally involved with that disease because 80% (800) of those with the gene develop the disease (Table 1)

Unfortunately, this table is not accurate enough to answer our questions precisely. Table 1 shows that it is possible to have the disease without the gene. Presumably this is because, in some individuals, an alternate set of sufficient conditions that does not involve the gene (environmental, etc.) has been met. We can infer, therefore, that some of those with the gene will also develop the disease, not due to the causal involvement of the gene, but rather because they too happen to have the alternate set of sufficient conditions. Of course, we can't know in individual cases which of the gene carriers has the disease because of the gene and which have the alternate sufficient conditions. But we can estimate their numbers by assuming

Table 1.

	Disease	No Disease	Total
Disease gene	800	200	1000
No disease gene	450	8550	9000
Total	1250	8750	10000

that approximately the same percentage of gene carriers will develop the disease because of alternate sufficient causes as amongst those without the gene. Of the 9,000 individuals without the gene, 450 still develop the disease, so we estimate 5% of individuals (with or without the gene) will get the disease without any causal action on the gene's part. In particular, 40 of the 800 people with both the gene and the disease need to be placed in a separate category (Table 2):

Using the data from Table 2, we can begin to answer these questions as follows:

1. The Global Question:

What percentage of the entire population will develop the disease in a way that casually involves the gene?

This would simply be the number of individuals where the gene was causally involved in the disease divided by the entire population, or $760/10,000 = 7.6\%$. This would tell us how large a problem the gene is causing in the population and thus, by extension, what could potentially be fixed by genetic manipulation. It does not really tell us anything interesting about whether the disease is genetic, however. We certainly would not want to say, for example, that rare diseases cannot be considered genetic (in fact, most paradigm examples of genetic disease are quite rare).

2. The Diagnostic Question:

What percentage of the subpopulation consisting of those with both the disease and the gene, developed the disease in a way that casually involves the gene?

In developing Table 2, we have already answered this question. We projected there that 5% of individuals, whether with the gene or without it, develop the disease for other reasons. The percentage of people in this subpopulation who do develop the disease in a way that causally involves the gene may be called the *Simple Etiologic Fraction* (SEF). In this case, the SEF would be the number of people where the gene was causally involved in the disease divided by the total number of people with the gene and the disease, or $760/800 = 95\%$. This would tell us what percentage of the population with the gene might potentially be impacted by therapeutic genetic manipulation.

It is tempting to view a high SEF (above 50%) as an indication that the disease is in fact genetic. However, we are tempted not because of what SEF actually indicates as what we might *think* it indicates. It does not really interest us to know that, in the subpopulation with *both* the gene and the disease, the gene is causally

Table 2.

	Disease	No disease	Total
Gene present, causally involved	760	0	760
Gene present, not causally involved	40	200	240
No disease gene	450	8,550	9,000
Total	1,250	8,750	10,000

Simple Etiologic Fraction $= 760/800 = 95\%$ *Population Etiologic* Fraction $= 760/1,250 = 61\%$
Attributable Risk $760/1,000 = 76\%$

involved in a high percentage of cases. It *would* interest us to know that the gene is usually causally involved amongst those with the disease *in general*, since this would allow us to discover whether the population causal intuition was met. But this is not what SEF tells us and we cannot derive that information from SEF (for this we need to answer the testing question). Simply put, SEF tells us nothing about the percentage of diseased individuals who owe their suffering to their genes. To do that, it would have to incorporate information about individuals with the disease, but who lack the gene (which it does not).

Similarly, SEF tells us nothing at all about the likelihood of developing the disease, given the gene (and thus cannot answer the prognosis question). We need this information to decide if the individual causal intuition is met. In order to do that, however, SEF would have to factor in information about people with the gene, but who remain disease-free (which it does not). Tempting as it might appear on first examination, SEF is pretty useless in answering the causal selection problem.

3. The Testing Question:

What percentage of the subpopulation having the disease developed the disease in a way that casually involves the gene?

This is asking for what epidemiologists call the *Population Etiologic Fraction* (PEF). In this case, it would be the number of individuals whose disease causally involved the gene, divided by the total number of diseased individuals, or $760/1250 = 61\%$. This does seem to be getting at something important concerning our intuitions about causal explanation. In particular, as long as the PEF > 50%, we know that *most* cases of disease in the population causally involve the genes. This is precisely the requirement of the population intuition, so it seems we need a stipulation in our concept of genetic disease that the PEF > 50%. We might be tempted to stop here and say that this is the only criterion for genetic status. However, although a high PEF insures that the population intuition is met, it does not assure us with respect to the individual intuition. Consider the following variation on our original case (Table 3).

Here, although it is quite true that most cases of disease in the population are caused by the genes (PEF $= 1,000/1,800 = 56\%$), it is not true that most of those with the gene have the disease in a way that causally involves the gene

Table 3.

	Disease	No disease	Total
Gene present, causally involved	1,000	0	1,000
Gene present, not causally involved	100	1,050	1,150
No disease gene	700	7,150	7,850
Total	1,800	8,200	10,000

Simple Etiologic Fraction $= 1,000/1,100 = 91\%$ *Population Etiologic* Fraction $= 1,000/1,800 = 56\%$
Attributable Risk $= 1,000/2,150 = 47\%$

(AR $= 1,000/2,150 = 47\%$). This violates our individual causal intuition and thus PEF needs to be supplemented as a concept of genetic disease.

4. The Prognosis Question:

What percentage of the subpopulation having the gene will develop the disease in a way that casually involves the gene?

Here we are asking for what epidemiologist call *Attributable Risk* (AR). In the original case from Table 2, we calculate AR by dividing the number of people whose disease causally involved the gene by the total number of people with the gene, or $760/1000 = 76\%$. Again, this does seem to be getting at something important in our concept of disease intuition. As long as the AR $> 50\%$, we know that *most* of those with the gene develop the disease through its causal involvement. This meets the requirement of the individual causal intuition, and thus we must also stipulate in our general account of disease that the AR $> 50\%$. Note, however, that just as a high PEF (answering the population intuition) does not guarantee a high AR, so a high AR (answering the individual intuition) does not guarantee a high PEF. Consider the following case (Table 4).

Here, although it is quite true that, in most cases, those with the disease gene develop the disease because of its causal involvement (AR $= 71\%$), it is also true that most cases of disease do not causally involve the gene (PEF $= 25\%$). This violates our population intuition and thus we cannot use the AR criterion alone.

7.4.3. Developing Specific Criteria

We can now get down to the business of applying the epidemiological concepts. Given the epidemiological outlook, what exactly constitutes a genetic disease? When do we have adequate grounds for saying that, although the causal system is very complex and neat necessary and/or sufficient conditions cannot be found, nevertheless we have identified a 'predominant cause' suitable to use in a simplified explanation? Clearly, any adequate answer to the causal selection problem must involve *both* PEF and AR, on pain of giving up one of our original intuitions. It remains an open issue, however, how *strongly* these should be interpreted. One

Table 4.

	Disease	No disease	Total
Gene present, causally involved	710	0	710
Gene present, not causally involved	90	200	290
No disease gene	2,000	7,000	9,000
Total	2,800	7,200	10,000

Simple Etiologic Fraction $= 710/800 = 89\%$ *Population Etiologic* Fraction $= 710/2,800 = 25\%$
Attributable Risk $= 710/1,000 = 71\%$

obvious possible answer would be to say that a disease is genetic whenever it is 'practically sufficient' within a given population:[22]

Practically Sufficient (PS): A disease is genetic whenever the gene's *Attributable Risk AND Population Etiologic Fraction* are both 100%. In plain English, this means that everyone in the population with the gene has the disease because of that gene's causal involvement AND that no one with the disease has the disease because of anything other than the gene.

This accords nicely with our intuitions that being a genetic disease has something important to do with sufficiency and necessity. On the other hand, PS does not make the mistake of claiming that the genes are either sufficient or necessary in the strict sense. It requires only that that the gene(s) be necessary components of each set of sufficient conditions *which occur in the population*. Since epidemiological analysis is relativized explicitly to some population of interest, it is entirely possible that what is practically sufficient in one population will not be in another. Practical sufficiency is thus the most minimal modification of strict necessity and sufficiency we can get away with and still do justice to highly complex causal systems.

We could cast the net of PS still wider by requiring not that AR and PEF both equal 100%, but rather that they come close – perhaps setting the threshold at 95%. Even here, though, we have a large problem. The main difficulty with PS is that, even though it is more practical and restricted than sufficiency and necessity in the strict sense, it is still too strong to apply to the vast majority of human diseases. This in itself would not be a problem, of course, but PS does not apply even to diseases where it seems we have excellent grounds for labeling them genetic. Its value for our purposes is thus mainly to anchor the endpoint of epidemiological concepts of disease – it represents as strong a notion of causation as it is possible to generate using population-relative epidemiological analysis of complex causal systems.

Recall that my goal is to develop a *minimally adequate* notion of genetic disease. What would a minimal threshold for PEF and AR be? Clearly, it would have to be at least 50%, as suggested by the use of the term 'most' in our original intuitions. Thus:

Minimally Epidemiological (ME): A disease is classified as genetic whenever both the Population Etiologic Fraction and the Attributable Risk exceed 50%.

If we refuse to sanction labeling a disease as genetic unless it has at least a 50% PEF and a 50% AR, then we have insured that both of our intuitions are met without adding any further restrictions. Most people with the disease will have the disease due to the causal involvement of the gene and most people with the gene will develop the disease because of the causal involvement of the gene. I have no principled objection to setting the thresholds at some number higher than 50%, but it will be difficult to defend any other number as more than a purely arbitrary selection. Moreover, it will be difficult enough to establish that current candidates for genetic disease status are legitimate in a minimal sense, much less a more restrictive one.

7.4.4. Applying the Account

I certainly do not wish to claim that all or even most of the diseases currently described as 'genetic' deserve that label. Indeed, it may well turn out that most diseases are so complex and with such high levels of variance that a causal selection in favor of any one type of factor (gene, environmental or whatever) is not justifiable. However, we can still clarify what the implications of the epidemiological approach are by attempting to at least fix the endpoints of the continuum.

On one end of the continuum we have diseases like obesity.[23] Obesity has increasingly been described as a genetic disease, since several genes have recently been discovered that regulate body weight, at least in mice. In this case, we can safely say that genes are 1) causally involved, 2) in principle manipulable and 3) practically sufficient to induce obesity in populations of lab mice on a normal diet. Even if we were talking about obesity in the general population of mice, however, we do not have the data we would need to claim that the condition is genetic in the ME sense, much less the PS sense. One reason for this, ironically, is that we do not know enough about *healthy* mice. That is, we do not know the prevalence of the 'obesity gene' among normal weight mice, and thus cannot accurately calculate the AR for the general population. We also do not know how many diseased mice lack the obesity gene, and thus cannot calculate PEF precisely either.[24]

Of course, we could *sample* normal weight mice and diseased mice and at least estimate these values in some rough sense, likely with enough accuracy to pass judgment on the disease's ME genetic status for populations of mice reared under 'normal' protocols (e.g., with diet and exercise held constant). However, in human populations, at least in the affluent West, it seems highly unlikely that the PEF for genes with respect to obesity will exceed 50%. Although it is certainly interesting to find that genes can induce obesity, these cases are likely only a small fraction of obesity cases. Other factors like diet (which cannot be controlled well in human populations, despite the best educational efforts of our medical community) will have extremely high PEF values. Diet will therefore almost certainly be singled out as the explanatory factor, *contra* genes, by any reasonable causal selection scheme. According to the epidemiological account, then, human obesity should not be classed as a genetic disease – genes do seem to play a role here, even an important role, but they are not the predominant cause of obesity.

At the other end of the continuum, a condition like Klinefelter's syndrome seems defensibly genetic in a strong sense.[25] Klinefelter's is caused by the presence of more than one copy of the X chromosome alongside a Y chromosome and results in numerous problems (e.g., abnormal sexual development, cognitive deficits, etc.). Here, the genes seem Practically Sufficient (PS) for the trait (AR and PEF of 100%). Again, of course, we will have difficulty supporting this claim as strongly as we would like, since we do not really know the relative numbers of people with the genetic anomaly who fail to exhibit the condition or who exhibit the condition for reasons other than their genes. However, in the case of entire additional chromosomes and complex symptomologies, we have very sound theoretical grounds to expect that very few people indeed will fall in these categories. It seems virtually certain then that Klinefelter's syndrome will qualify as genetic in the ME sense, and it's an excellent bet that it will in the PS sense as well, so it's a good choice for the other end of the continuum.

Cystic fibrosis (CF) is a much more complex case. CF is associated with any of at least 300 different known mutations in the Cystic Fibrosis Transmembrane Conductance Regulator (CFTR) gene. There is no clear relationship between the severity of symptoms and specific mutations, and there are even cases where individuals with a CFTR mutation do not have the disease. CF thus could not be classified as genetic in the PS sense. Even on the ME account, the case is not perfectly clear. The PEF for the gene (collectively, though not for any single mutation) is probably at or close to 100% – we have, at least arguably, decent data here since it is not unusual for CF sufferers to be tested for the gene. Again, however, the AR is simply not known, though we do know it is definitely < 100%, since we know of individuals with the mutations who do not have the disease. There are even suggestions that CF mutations may confer an advantage on their (heterozygote) carriers by increasing the resistance to lung infections. So we have both empirical and theoretical reasons to be cautious when it comes to estimating the AR for CF. However, I would argue that it still seems a good bet (though not an excellent one) that CF will turn out to be genetic in the ME sense, given the fundamental function of the CFTR receptor in the cell, that individuals with the mutations who manage to function normally seem to be relatively rare (and thus that AR is fairly high). It helps that such individuals need not be terribly rare since, unless these lucky carriers actually *outnumber* those with the gene and the disease, CF will still qualify as genetic in the ME sense (AR > 50%).

7.4.5. Difficulties

7.4.5.1. Population Relativity   The epidemiological account is able to handle the complex and highly variable causal systems we know to be operating in human disease without engaging in damaging oversimplification and while also restricting itself to questions of a decidable empirical nature. However, this advantageous arrangement does come at a price – as an inherently statistical account, any epidemiological explanation of a disease must be explicitly relativized to some particular population. This means there is no guarantee that the explanation for a trait in one population will hold true for other populations – indeed, there is often excellent reason to think that it will not. For example, an epidemiological analysis would show that lactose intolerance is a genetic trait for most populations in Western Europe and the U.S. (where consumption of milk products is common but the gene for intolerance is not), yet it is clearly environmental in many Asian countries (where the genes for intolerance are common but consumption of milk products is not) (See also Burian, 1981–82, Stern, 1973 and Smith, 1992).

Another counterintuitive possibility is suggested by the following example:[26] suppose there is a village somewhere in Africa where everyone routinely drinks from a contaminated water supply. As a result, most of the villagers suffer from a nasty parasitic infestation. On closer examination, it turns out that those who remain parasite free do so because the genes they possess confer immunity. The epidemiological account would conclude that, in this village, the parasitic infection is genetic because genes are practically sufficient to bring about the affliction. Drinking contaminated water, strangely enough, is not selected as the explanatory cause because both the etiologic fraction and the negative etiologic fraction of the water are 100% (and thus the etiologic residue is 0).

What exactly is going on here? First, there may be some confusion about casual selection vs. causal connection questions. Denying that one should select the contaminated water as explanatory doesn't in any way diminish the importance of its causal role. Indeed, no one would deny that, in this case, drinking the water is a necessary condition for the trait (and even that cleaning up the water supply might be the most efficient means of preventing future outbreaks). However, this response still doesn't seem to completely allay our misgivings.

The other reason the conclusion seems peculiar is that the example specifies a very restricted population – that of the village alone. Yet we do not normally have such a restricted population in mind when we ask typical causal questions like, 'Why do *people* (not villagers in Ngome) get this condition?' Rather, we intuitively evaluate causal selections in terms of the more inclusive populations with which we are more comfortable (e.g., humans in general, citizens of the affluent West like ourselves, etc.). Thus, it seems silly to conclude that the condition is genetic because we know we would never accept such a conclusion about these 'default populations.' But the epidemiological account would not force us to do that – indeed, the condition would be explained as the result of contaminated water for the vast majority of populations in the modern world.

I freely grant that this is not the way we ordinarily think of disease. However, we should not forget that the root cause of this problem is the fact that human diseases are so complex and human populations so varied that monocausal explanations are necessarily distorted. The epidemiological account does not create this problem, it simply makes it very clear when the explanatory costs of simplification outweigh the practical benefits of brevity. In other words, the messenger should not be blamed for the message

Nevertheless, the population relativity is especially frustrating in situations where it is not clear precisely how one is to choose the population in question. Ultimately, there can be no solution to this other than a pragmatic one – the population chosen must be one of sufficient interest to one's audience. It is true that, given our different interests, such a procedure will inevitably result in a great many competing explanations for the same trait. However, it is also true that we very often *will* have a clearly defined population in mind when we create an explanation. The United States Department of Health, for example, is primarily concerned with the explanation of diseases *within the U.S. population*. Those still worried about relativism creeping into our explanations can take solace from the fact that the population in question must at least be cited explicitly and the rules for what counts as adequate within any population are well-defined. The epidemiological account does do some violence to our intuitive demand for a population independent account of disease, but it hardly constitutes an 'anything goes' abandonment of objectivity.

7.4.5.2. Individual cases   It also might plausibly be objected that the epidemiological account of explanation is problematic because it's completely at a loss when faced with the task of explaining *individual* occurrences of a trait.[27] For example, if a patient insists on knowing what caused *his particular* case of cancer, it does not help him very much to cite the relative prevalence of causal factors within the larger population of which he is a part. This is indeed counter-intuitive, given the focus of modern medicine (especially in the United States) on the care of *individual* patients. However, before taking this criticism too far, we should consider whether or not there really is an *alternative*.

I ask the reader to indulge me for a moment in a hypothetical example. Suppose we have an alien physician who comes to Earth with such advanced technology that he knows the complete causal connection picture. He abducts a cancer patient and does a compete exam, determining everything there is to know about this particular case. However, since he only has the one abductee, he does not know anything about the more general human population. Suppose the patient asks him if *his particular case* of cancer is genetic. If the alien really wanted to answer the patient's question, he would have only three possible routes. First, he could simply indicate the entire causal matrix for that patient (in effect answering the causal connection question but saying nothing about causal selection). Second, he could engage in so-called *counterfactual analysis*, where one tries to make projections about what *would have* happened in a particular case *were* the circumstances different from what they actually are/were. However, most of these projections would be mere guesswork unless the alien is allowed to learn more about how the causal connection picture in this individual case differs from that in other cases. Thus, counterfactual analysis relies on population level information. Besides, it's not clear exactly how this sort of exercise will address the patient's very practical question.

The only other option left to the alien doctor would be to explicitly import population data. In a population, but not in an individual, there will be *variation* in causal factors. In a large enough population, one can find almost any combination of relevant casual (risk) factors. By crunching these numbers, we can calculate the actual risk of a particular disease, given any set of initial conditions. In other words, population data would allow us to solve the problem – *but only by cheating and bringing in information from outside the individual case.* We simply cannot say much, if anything, about the relative importance of causal factors if we rely only on the perspective of the individual patient – the causal selection problem *requires* data from populations for its resolution.

To suggest, therefore, that reliance on populations is somehow a difficulty is to imply endorsement of an impossible alternative. Without some, at least implicit, appeal to population-level information such as which factors vary and in what way, there are simply no grounds for causal selection, no matter what account one favors. Accounts other than the epidemiological one will still use population information because they must, but they sneak it in the back door without ever being clear about what they are doing. The result is a fuzzy and misleading analysis of the problem. Seen in this light, the epidemiological requirement that the explanation be relativized explicitly to a carefully delimited population is a virtue, not a vice.

7.5. CONCLUDING REMARKS

The concept of a genetic disease is neither well-developed nor generally defensible as it is employed in the literature. There are very few attempts to make the criteria for causal selection explicit in general, and almost none in the specific case of human disease. One must reason backwards from the kinds of claims one finds in the literature to implicit notions of causal selection, but the notions thus uncovered are not able to withstand the harsh light of critical scrutiny. Rather, they owe their survival to a combination of an uncritical atmosphere and a very ambiguous formulation.

The epidemiological account of genetic traits is an analysis of genetic disease which is both practical and theoretically defensible. This account avoids many of the criticisms leveled against its rivals, while still preserving a use of 'genetic trait' which is useful and informative. However, we must apply it with great care. In particular, we must always keep in mind that a great many traits, likely even a large majority, will not meet *any* defensible criterion of genetic status. This is not a failure of the epidemiological account, so much as an admission that the world of biological causation is far too complex and varied to admit of simplistic categorization.

Department of Philosophy and Religion, Clemson University, Clemson
South Carolina

NOTES

[1]   Some exceptions include Burain 1981-2, Hesslow 1984, Norell 1984, Moss 1992, and Magnus 1996.
[2]   It is not necessary for our purposes at the moment to differentiate between variability between individuals in a population and variability within an individual over time, though both are typically operative in biological contexts.
[3]   This is a slightly more inclusive version of the distinction introduced by Hesslow (1988).
[4]   Most scientists resist any suggestion that their work is, in any meaningful way, metaphorical. An interesting discussion by a biologist of the use of metaphors in precisely these situations can be found in Nijhout (1990).
[5]   Throughout the paper, I use 'gene' and 'genes' in a rather loose way to refer to all those elements of DNA (genes, alleles, regulatory sequences, etc.) which causally impact the trait. This choice itself is, of course, a necessary simplification of a very complex situation.
[6]   Gene-centrism has come under attack of late by a number of scholars (Lewontin 1984; Levins and Lewontin 1985; Oyama 1985; Griffiths and Gray 1994; van der Weele 1995; Griffiths 1996; Gilbert et al., 1996; Sterelny, et al. 1996). It is actually possible to distinguish at least two distinct flavors of gene-centrism: *genic selectionism* is the view that the process of natural selection is best understood as occurring at the level of the genes while *the genetic program view* is the notion that organismal development and cellular functioning are controlled by a program on information contained in the genes (Smith 1994).
[7]   Throughout the paper, I will use the term 'epigene' as a grab-bag term covering anything which is not physically contiguous with nucleic acids (e.g., proteins, environmental factors, etc.).

[8] Even if we risk the accusation of an ad hoc move and *define* life as beginning with the advent of nucleic acids, the existence of 'genes' seems to imply an appropriate epigenetic environment in which the sequence of bases on the nucleic acids is translated into useful information.

[9] For a discussion of some of these points see Wilson 1977, Brandon 1982 and Smith 1994.

[10] It frequently seems, as Gould and Lewontin (1979) observe in a different context, as if the strategy is to admit the phenomena (epigenetic influence), yet restrict its domain of application so narrowly as to be of only academic interest (see also Wimsatt 1980).

[11] Of use, that is, in illuminating the *causal* nature of the trait. One could argue, of course, that such descriptions are highly useful for other purposes – securing funding for genetic research, etc.

[12] To give just one example, human pulmonary morphology involves a process which creates and then reabsorbs embryonic gill slits – hardly a model of directness, but quite reliable.

[13] This phenomenon has been known since the earliest days of genetics. For example, the *bithorax* mutant in the fruitfly can be produced either with a genetic mutation or by exposing flies to ether, the latter method producing a 'phenocopy' of the genetic mutation.

[14] Hart and Honore themselves suggest that their account might be altogether too subjective for use in the sciences.

[15] Precipitating factor analysis is even used by the World Health Organization as one of the two criteria by which physicians should select the cause of death, though it seems ill suited to such contexts (Lindahl 1984).

[16] Russell (1916) points out that there will *always* be interceding events between any putative cause and its effect (see also Brandon, et al. 1994).

[17] One theory of cancer paints precisely this picture – mutations and precancerous growth are universal, though an environment conducive to tumor formation (compromised immune system, etc.) is not.

[18] It may be a bit unfair to show Koch as requiring *strict* sufficiency and necessity through the use of 'always' and 'never' in the postulates. Whatever the historical accuracy of this move, it serves to illustrate the poverty of such an approach when applied to complex diseases. At a minimum, Koch's approach would have to be supplemented with another technique (which he does not provide) on pain of leaving the causal selection to the whim of the researcher.

[19] Indeed, exposure to the TB bacillus used to be essentially universal before modern public health measures were instituted in the Western world. Today, exposure to the bacillus tends to be identified as the cause of TB, but in the 1940's it would be more accurate to identify differences in immunity rather than exposure as the culprit (Stern 1973).

[20] In fact, the reference population must have at least two subpopulations *differing* in the trait in question. This is because the phenomenon actually being explained is the *variance* in the trait between the two subpopulations, not the trait simplicitur.

[21] This is quite difficult to do consistently – an earlier version of these questions suffered from precisely this sort of confusing language (Smith 2001)

[22] See also Gifford 1990 and Wulff 1984b for similar attempts to develop a notion of practical sufficiency.

[23] Of course, it is an open question as to whether a condition like obesity should really qualify as a disease. It is not my intent to offer a general account of disease, so I will simply assume for the purposes of argument here that it does.

[24] In fact, this sort of situation poses a major epistemic problem for any theoretically adequate notion of disease causation. Studies to determine the actual incidence of disease genes in healthy populations will be extremely expensive and may not be available for many years to come.

[25] It has been suggested that perhaps Klinefelter's is not a genetic disease at all, since it is not heritable. However, it seems more accurate to say that heritability is an *indicator* of a disease's genetic status rather than a necessary condition.

[26] I am indebted to David Magnus for this example.

[27] This is because there is no variation within an individual with respect to the causal factors to be analyzed.

REFERENCES

Allen G (1980) Dialectical materialism in modern biology. Science and nature 3 3:43–57

Brandon RN (1982) The Levels of Selection. In: Asquith P and Giere R (eds) Proceedings of the philosophy of science association 1980, 1:315–323

—— (1990) Adaptation and the environment. Princeton University Press, Princeton

Brandon RN, Antonovics J, Burian R, Carson S, Cooper G, Davies PS, Horvath C, Mishler B, Richardson RC, Smith K, Thrall P (1994) Sober on Brandon on screening-off and the levels of selection. Philos Sci 61(3):475–486

Burian R (1981–82) Human sociobiology and genetic determinism. The philosophy forum XIII(2–3): 43–66

Collingwood RG (1938) On the so-called idea of causation. Proceedings of the Aristotelian society pp 85–112

Dawkins R (1976) The selfish gene. Oxford University Press, Oxford

Ducasse CJ (1924) Causation and the types of necessity. University of Washington publications in the social sciences 1(2):69–200

Fox SW, Dose K (1972) Molecular evolution and the origin of life. San Francisco, Freeman

Gardenfors P (1980) A pragmatic theory of explanation. Philosophy of science 47:404–423

Gifford F (1990) Genetic traits. Biology and philosophy 5(3):327–347

Gilbert SF (1997) Developmental biology, 5th edn. Sinauer Associates Inc, Sunderland, MA

Gilbert SF, Opitz JM, Raff RA (1996) Resynthesizing evolutionary and developmental biology. Dev Biol 173(2):357 ff

Gould SJ, Lewontin RC (1979) The spandrels of San Marco and the Panglossian Paradigm: a citique of the adaptationist program. Proceedings of the Royal Society of London, B205:581–598

Griesemer J (1994) The informational gene and the substantial body: on the generalization of evolutionary theory by abstraction. In: Cartwright N, Jones M (eds) Varieties of idealization, Poznan studies in the philosophy of science and humanities, Radpi Publishers, Amsterdam

Griffiths PE, Gray R (1994) Developmental systems and evolutionary explanation. J Philos 91:277–304

Griffiths PE (1996) Darwinism, process structuralism and natural kinds. Philos Sci 63(3) Supplement: S1 ff

Hart HLA, Honore AM (1959) Causation in the Law. Oxford University Press, Oxford

Hesslow G (1983) Explaining differences and Weighting Causes. Theoria 49:87–111

—— (1984) What is a genetic disease? on the relative importance of causes. In: Nordenfelt L, Lindahl BIB (eds) Health, disease, and causal explanations in medicine, Dordrecht, Reidel

—— (1988) The problem of causal selection. In: Hilton DJ (ed) Contemporary science and natural explanation: commonsense conceptions of causality, NYU Press, New York

Hilton, DJ (1988) The problem of causal selection. In: Hilton DJ (ed) Contemporary science and natural explanation: commonsense conceptions of causality, NYU Press, New York

Hull D (1981) Units of evolution: a metaphysical essay. In: Jensen UL, Harre R (eds) The philosophy of evolution, Harvester Press, Brighton

Kauffman SA (1993) The origins of order. Oxford University Press, Oxford

Levins R, Lewontin RC (1985) The dialectical biologist. Harvard University Press, Cambridge

Lewontin RC (1984) Not in our genes: biology, ideology and human nature. Pantheon Books, New York

Lindahl BIB (1984) On the selection of causes of death: an analysis of WHO's rules for selection of the underlying cause of death. In: Nordenfelt L, Lindahl BIB (eds) Health, disease, and causal explanations in medicine, Dordrecht, Reidel

Mackie JL (1965) Causes and conditions. American Philosophical Quarterly 2(4):245–264

—— (1974) Cement of the universe. Clarendon Press, Oxford

Magnus D (1996) Gene therapy and the concept of a genetic disease [on-line] Genetics and ethics, a global conversation, University of Pennsylvania Center for Bioethics wwwmed.upenn.edu/bioethics/

Martin R (1978) Judgments of contributory causes and objectivity. Philos Soc Sci 8:173–186

Mayr E (1983) How to carry out the adaptationist program? Am Nat 121:324–334

—— (1988) Towards a new philosophy of biology: observations of an evolutionist. Belknap Press, Cambridge

Mill JS (1859) A system of logic. Longmanns, London (reprinted 1961)

Moss L (1992) A kernel of truth: on the reality of the genetic program. In: Proceedings of the philosophy of science association 1992, 1:335–348

Nagel E (1961) The structure of science: problems in the logic of scientific explanation. HBJ, New York

Nijhout HF (1990) Metaphors and the role of genes in development. Bioessays, 12(9):441–446

Norell S (1984) Models of causation in epidemiology. In Nordenfelt L, Lindahl BIB (eds) Health, disease, and causal explanations in medicine, Dordrecht, Reidel

Oparin AI (1924) The origin of life. Dover, London (reprinted 1953)

Oyama S (1985) The ontogeny of information. Cambridge University Press, Cambridge

Rosenbaum M, Leibel RL, Hirsch J (1997) Obesity. N Engl J Med, 337(6):396–407

Rothman KJ (1976) Causes in American Journal of Epidemiology, 104:587–592

Russell B (1916) Address to the Aristotelian society, as reprinted in Russell B (ed) Human knowledge: its scope and limits. George, Allen & Unwin, London

Ryle G (1949) The concept of mind. Hutchinson, London

Salmon WC (1984) Scientific explanation and the causal structure of the world. Princeton University Press, Princeton

Smith K C (1992) The new problem of genetics: a response to Gifford. Biol Philos 7:331–348

—— (1993) Neo-Rationalism vs. Neo-Darwinism: integrating development and evolution. Biol Philos 7(4):431–451

—— (1998) Equivocal notions of accuracy and genetic screening of the general population. The Mount Sinai journal of medicine 65(3):178–183

—— (2001) A disease by any other name. The European journal of medicine, philosophy and health care 4(1):19–30

Sterelny K, Smith KC, Dickison M (1996) The extended replicator. Biol Philos 11:377–403

Stern C (1973) Principles of human genetics. Freeman, San Francisco

Van der Weele C (1995) Images of development: environmental causes in ontogeny. Dissertation, Amsterdam, Vrije Universiteit

Van Fraassen BC (1980) The scientific image. Oxford University Press, Oxford

Walters L, Palmer J (1997) The ethics of human gene therapy. Oxford University Press, New York

Williams GC (1966) Adaptation and natural selection. Princeton University Press, Princeton

Wilson DS (1977) Structured demes and the evolution of group-advantageous traits. Am Nat 111: 157–185

Wimsatt WC (1980) Reductionistic research strategies and their biases in the units of selection controversy. In: Nickles T (ed) Scientific discovery: case studies, Dordrecht, Reidel pp213–259

Wulff HR (1984) The causal basis of the current disease classification. In: Nordenfelt L, Lindahl BIB (eds) Health, disease, and causal explanations in medicine, Dordrecht, Reidel

8. WHY DISEASE PERSISTS: AN EVOLUTIONARY NOSOLOGY

8.1. INTRODUCTION

At the end of *On the Origin of Species*, Darwin wrote, 'as natural selection works solely by and for the good of each being, all corporeal and mental endowments will tend to progress towards perfection' (Darwin, 1859, p. 489). Taken at face value, this sentence would seem to imply that natural selection will lead to the amelioration, if not the elimination, of disease. And yet, after thousands of generations of human evolution, and thousands of millennia of evolution by natural selection before the appearance of Homo sapiens, disease persists. An evolutionary nosology, a classification of the evolutionary causes of disease, may help to rationalize the persistence of disease and the inability of natural selection to eliminate it (Nesse & Williams, 1994).

The Darwinian theory of evolution by natural selection is based on the recognition that populations of organisms exhibit heritable variation in traits that are associated with survival and reproductive success, or fitness. The differential survival and reproduction of organisms—their differential mortality and fertility rates—leads to the differential transmission of alleles from one generation to the next. Evolution is commonly thought of as the changes in allele and genotype frequencies—and the accompanying change in the distribution of phenotypes in a population—that result from this differential transmission of alleles. Genetic (and phenotypic) variation is thus central to our understanding of the evolutionary process. Moreover, the properties of organisms develop and change over time, as they progress through their life course. Species comprise populations of organisms that share a common evolutionary heritage and whose members retain the ability to reproduce with one another, but which otherwise are characterized by variation and change. Boorse (1977) has argued that health is a value-free, statistical concept, based on a 'species-typical' level of physiological function. As Kovács (1998) and Nesse (2001) have pointed out, however, there are no objective criteria that distinguish 'normal' from 'diseased'; the delineation of this boundary must entail value judgments. Indeed, any essentialist notion of health and disease seems incompatible with an evolutionary perspective, and an appreciation of the importance of variation in biology. Despite the problem of delineating health from disease, disease is generally understood to involve suffering, decreased function, limitations in achieving one's goals, and perhaps an increased likelihood of death. Indeed, a more precise definition of disease may neither be possible nor necessary (Hesslow, 1993).

Diseases are like species (Faber, 1930); like species, they are best understood from what Ernst Mayr (1964) has called a 'population perspective.' Diseases are

111

H. Kincaid and J. McKitrick (eds.), Establishing Medical Reality, 111–121.
© 2007 *Springer*.

groupings of the more-or-less similar life histories of individual diseased persons. Diseased people have altered ways of living (ways of living manifested by suffering, decreased function, etc.); but just as individual people differ from one another in health, they differ from one another in the ways in which diseases affect their lives. Moreover, diseased individuals go through what may be considered the life course of their disease. The manifestations of disease appear, change over time, and may eventually disappear or may remain until the diseased individuals die. Thus diseases, like species, are characterized by variation and change. The classification of diseases, like the distinction between health and disease, entails value judgments, and depends on the purposes to which the classification is put. Historically, diseases were defined and classified by their clinical manifestations, the signs and symptoms of diseased people. Now, we increasingly rely on laboratory criteria and define diseases according to their causes (Cunningham, 1992). As Rees (2002) has noted, for physicians, causal selection is often pragmatic: 'The cause of disease is not ... some objective God's eye summary of pathophysiology, but rather an operational statement of where we think the Achilles' heel of a disease might be.' This essay focuses on a different set of causes, the phylogenetic or ultimate causes—the reasons why natural selection has not eliminated diseases.

8.2. NATURAL SELECTION IS NOT THE ONLY EVOLUTIONARY PROCESS

As noted above, evolution—or microevolution—is often defined as a change in allele or genotype frequencies in a population over time. Natural selection is one of the mechanisms that can change allele frequencies, and is the only process that can lead to adaptations, to increases in the mean fitness of populations. Natural selection will act to reduce disease by eliminating alleles that are associated with infertility and premature death (death before the end of the period of reproduction and child-rearing). But natural selection is not the only evolutionary force—other processes, including mutation, genetic drift, gene flow, and gametic selection, can also change allele frequencies, and can counter the effects of natural selection. Consider mutation. Mutation is an important source of novelty and as such is an essential component of the evolutionary process. Because human beings are complex, well-integrated organisms, however, most mutations that affect the structure or abundance of proteins are likely to be deleterious. These deleterious mutations will be maintained at low frequencies, frequencies determined by mutation-selection balance; in a steady state, the rate at which deleterious, disease-associated alleles arise by mutation equals the rate at which they are removed by natural selection. The removal of deleterious mutations is as important a component of natural selection as is the preservation of favorable mutations. Although individual single-gene Mendelian diseases are rare, together they cause a significant burden of disease (OMIM, 2006).

Different human populations evolved in different environments, and the individuals in these populations are adapted to the environments in which their ancestors evolved. Migration, or gene flow, brings people and their genes into new populations, and is another important source of genetic novelty. If this migration brings people to new environments, however, it may lead to disease. The skin cancers that develop in fair-skinned people who travel or move to the tropics exemplify diseases that result from gene flow.

8.3. PLEIOTROPY, EPISTASIS AND LINKAGE

Although evolution depends upon heritable variation in traits that affect fitness, most traits do not exhibit a simple relationship between genotype and phenotype. Many genes are pleiotropic—that is, they affect more than one phenotypic trait. Pleiotropic genes may have both beneficial and deleterious effects; as long as the beneficial effects balance the harmful ones, these alleles will be maintained in a population. The globin genes are good examples of pleiotropic genes. Two globin genes, α and β, are expressed at high levels after the neonatal period. Their gene products, the α and β chains of hemoglobin, affect (among other traits) the traits of oxygen transport and susceptibility to malaria. The Hb S allele of the β globin locus is maintained in populations in malarious regions because it increases resistance to malaria, even though, when homozygous, it results in sickle cell anemia and has deleterious effects on red blood cell survival and oxygen transport. Whether considered from the perspective of pleiotropy or of heterozygote advantage, Hb S provides a model for understanding how alleles with deleterious effects may be maintained in populations. A new allele will increase in frequency until its deleterious effects balance its beneficial ones; in other words, until the mean fitness of genotypes containing the allele equals the mean fitness of the genotypes without it. Most genes are pleiotropic, and many alleles are associated with increased risk of disease. If these alleles also have beneficial effects, they—and their associated diseases—will be maintained in the human population.

Not only do individual genes affect many traits, but many genes may interact to affect a single trait. The non-additive effects of genes—or, more properly, gene products—on fitness is known as epistasis. If genes interact epistatically, then the fitness effects of alleles at one locus may depend on the specific alleles that are present at a second locus; in other words, different combinations of genotypes may optimize fitness. The growing appreciation of the phenotypic diversity of genetic diseases has led to increased interest in modifier loci, loci that affect the phenotypic consequences of disease-associated genes. As disease-associated alleles (Hb S, for example) spread in a population, there will be selection for mutations at other loci that decrease the severity of the diseases associated with these alleles. Mutations that result in the persistence of fetal hemoglobin are one class of mutations that decrease the severity of sickle cell anemia; not surprisingly, these mutations have spread in populations that have a high prevalence of Hb S.

Hereditary persistence of fetal hemoglobin by itself has little physiological conse-quence, and is not considered a disease. Mutations at other modifier loci, however, may themselves be associated with disease. One well-studied example of epistatic interactions affecting human disease concerns the interactions between mutations that regulate the production of the α and β globin chains. Mutations that decrease production of either globin chain can result in diseases that are known generically as thalassemias. Thalassemias are among the most common genetic disorders. Thalassemia mutations, like Hb S, have been maintained at high frequencies in some populations because, when present in heterozygous form, they confer resistance to malaria. When present in homozygous form, however, both α-thalassemia and β-thalassemia alleles may cause severe, often fatal disease. These diseases appear to result from the unbalanced production of the two globin chains, rather than from the deficient production of one. Thus, the presence of an α-thalassemia mutation, which reduces α chain synthesis, may ameliorate the severity of homozygous β-thalassemia (Weatherall, 2001). The details of this interaction are complex, because of the diversity of thalassemia mutations, the expression of other globin genes, and the selective effect of malaria. Nonetheless, the principle is straightforward: α-thalassemia mutations may increase fitness in people with β-thalassemia mutations, but decrease fitness in people who have normal rates of β globin synthesis. As a result of this epistatic interaction, α-thalassemia alleles may spread in populations that have a high incidence of β-thalassemia. Even if alleles have deleterious, disease-associated effects when present in people with some genotypes, they may be maintained in a population because of their beneficial effects in people with other genotypes. Again, these disease-associated alleles will be maintained at frequencies at which the mean fitness of genotypes bearing the alleles equals the mean fitness of genotypes without them.

Genetic linkage is another mechanism that may hinder the ability of natural selection to remove disease-associated mutations. When two genetic loci are tightly linked, so that recombination between them is rare, they behave as a single pleiotropic gene. Under these conditions, alleles at these loci will change in frequency according to the balance between their combined beneficial and deleterious effects. For this reason, even deleterious alleles may increase in frequency if they are tightly linked to beneficial alleles. The increase in frequency of an allele because of linkage to a beneficial allele is known as hitchhiking. The allele that is responsible for most cases of hemochromatosis in Europe may have spread by hitchhiking, because of its linkage to specific HLA alleles (Distante et al., 2004).

8.4. LIFE HISTORY STRATEGIES, AGING AND DEVELOPMENTAL PLASTICITY

Natural selection increases the mean fitness—or, more correctly, the mean inclusive fitness—of populations. Fitness, however, is not the same as absence of disease. Fitness is success in producing progeny who are themselves reproductively successful; in genetic terms, fitness is success in contributing genes to the population gene pool. Although survival is an important component of fitness, fitness entails survival only through the age at which individuals reproduce or contribute to the survival and reproductive success of their offspring. Aging provides perhaps the best example of the distinction between fitness and the absence of disease. Aging may be defined as 'a progressive, generalized, impairment of function resulting in a loss of adaptive response to stress and an increasing probability of death,' and frequently accompanied by a decline in fertility (Kirkwood, 1999). In nature, most organisms do not die of 'old age'; they die of other, 'extrinsic' causes (accidents, predators, starvation, etc.—causes that cannot be eliminated by natural selection) before they have a chance to age. Because fewer and fewer individuals survive to older and older ages, the intensity of natural selection will decline with age. The intuitive basis of this relationship was stated clearly by Charlesworth (1994, p. 197):

[U]navoidable sources of mortality cause the size of a cohort to dwindle with advancing age, so that a gene with delayed expression will have a smaller net effect on the composition of a population than a gene which is expressed early in life.

The precise age-dependence of the force of natural selection is complicated, because it depends on population growth rates, the different reproductive histories of men and women, and the post-reproductive contributions of parents to the fitness of their children. Nonetheless, the shape of this relationship, and its consequences, are clear: the force of natural selection, its ability to eliminate alleles that are associated with disease and that lead to death at specific ages, is high until the onset of reproduction, and then declines. Aging is a consequence of this decreasing power of natural selection.

Humans, like other organisms, have or can acquire only finite resources, and they must allocate these resources—including time, which is perhaps our most precious resource—to some combination of growth, somatic maintenance, and reproduction (Hill & Kaplan, 1999). Moreover, they must adjust this allocation over their life course, as they space their reproductive effort. Natural selection is expected to optimize this resource allocation in ways that maximize individuals' inclusive fitness (Hill & Kaplan, 1999). Because resources are finite, however, and some must be devoted to growth and reproduction, somatic maintenance is necessarily imperfect. As a result, mutations go uncorrected, abnormally folded proteins accumulate, cells die, and we age.

Williams (1957) proposed that aging resulted from 'antagonistic pleiotropy'; alleles that promote reproduction early in life will spread in populations, even if they result in aging and death, because their early benefits outweigh their later deleterious effects. The 'disposable soma' theory of aging is a synthesis of the concept of antagonistic pleiotropy with that of the allocation of finite resources. The disposable soma theory proposes that the pleiotropic genes which result in aging are genes that divert resources from somatic repair to reproduction; again, these genes will spread because they increase fitness, even though they may be associated with disease later in life (Kirkwood, 1999).

Natural selection will not necessarily lead to a fixed, genetically determined allocation of resources between growth, somatic maintenance, and reproduction; indeed, it may be advantageous for organisms to vary this allocation in response to their own environment and individual condition. As Hill and Kaplan (1999) note, phenotypic plasticity evolves 'because the optimal phenotype varies with conditions, and genetic variants coding for the ability to modify phenotype adaptively sometimes can out compete variants that produce the same phenotype in all environments.' The cues that developing organisms use to adjust their resource allocation and reproductive schedule may be nutritional or psychosocial. Individuals in a stable, resource-rich environment may optimize their fitness by postponing and limiting their reproduction, and investing heavily in their own somatic maintenance and in their children. In contrast, individuals in an unstable or resource-poor environment are likely to reproduce earlier and have more children, even if doing so increases the risk of disease and compromises their longevity (Coall & Chisholm, 2003). Barker and his colleagues have shown that fetal nutrition has long-term consequences for adult health (Barker, 1992). In particular, low birth weight, or in utero growth retardation, appears to predispose people to develop hypertension and insulin resistance later in life. According to the 'thrifty phenotype' hypothesis, fetuses and newborn infants respond to a poor nutritional environment in ways that improve their chances of survival to the age of reproduction, even at the expense of risking disease later in life (Hales and Barker, 2001). Although the specific genes and physiological pathways involved in the thrifty phenotype response have yet to be elucidated, this phenotypic plasticity appears to be an evolutionary adaptation that enables developing organisms to respond appropriately and maximize their fitness in diverse nutritional environments.

Psychosocial or socioeconomic cues may also influence the allocation of resources between somatic maintenance and reproduction. People of low socioeconomic status, like those with poor fetal growth, age earlier and have a decreased life expectancy (Marmot, 2003; Wilkinson, 1997). The increased burden of disease seen in people of low socioeconomic status, like that in people of low birth weight, may reflect their phenotypic plasticity and their preferential allocation of resources away from long-term somatic maintenance toward short-term survival and reproduction (Coall & Chisholm, 2003). Again, this phenotypic plasticity is itself an evolutionary adaptation, the result of natural selection. Amelioration of the diseases associated with poor fetal growth and with low socioeconomic status will require improvements in fetal nutrition and reductions in socioeconomic differentials.

8.5. LEVELS OF SELECTION

We focus on natural selection acting on the differential survival and reproductive success of organisms, in part because we are organisms and in part because this is where natural selection appears to be most important. But natural selection acts at many levels of biological organization, on any entities that can count as individuals (Hull, 1980). These entities may include, in addition to organisms, DNA sequences, gametes, cells, and groups. A large fraction of the human genome consists of repetitive DNA sequences, sequences that survive and reproduce within the environment of our genomes (Doolitte & Sapienza, 1980; Orgel & Crick, 1980). For the most part, this 'selfish DNA' does not cause disease—evidently, natural selection has led to the elimination of those repetitive DNA sequences that do. On the other hand, the proliferation of trinucleotide repeat sequences and of transposable repetitive sequences may lead to disease. Gametic selection is a process that leads to the spread of alleles because of the preferential survival of gametes containing these alleles. This process has been described in mice, in which it is associated with male infertility, but it hasn't yet been reported in humans.

How different 'levels of selection' interact in evolution is still a matter of controversy (Sterelny & Griffiths, 1999). Cancer may be a good example of natural selection acting on cells, at the expense of the organisms of which these cells are a part. Multicellular organisms provide an environment in which their component cells can grow and replicate. In animals, which have distinct but genetically identical germ cells and somatic cells, survival and reproduction has entailed the evolution of mechanisms that limit the reproduction of somatic cells (Buss, 1987). Nevertheless, there will always be selection for cells that escape these constraints, either by genetic mutation or by epigenetic change, and that can replicate and spread within the organism. Cancer may be understood as the unfortunate outcome of selection at different levels of biological organization—selection of mechanisms that constrain the growth of somatic cells and selection of cells that escape these constraints (Greaves, 2002). The mechanisms that prevent the unrestrained growth of somatic cells are so effective that the development of cancer requires several rounds of mutation and selection. The most common cancers are cancers of epithelial tissues—lung, colon, and breast—that replicate throughout life. The risk of developing cancer is a trade-off for the benefits of epithelial regeneration. Lymphomas and leukemias are frequently associated with chromosomal breaks and translocations. These tumors result from the inappropriate activity of the enzymes that promote genetic recombination of the antibody and T-cell receptor genes during lymphocyte maturation; the risk of developing these tumors may be seen as a trade-off for the benefits of adaptive immunity.

Individual organisms contain multiple genomes—mitochondrial and nuclear genomes, or maternal and fetal genomes. For the most part, these genomes interact co-operatively, or symbiotically, in development. Nonetheless, the action of natural selection on these different genomes may lead to disease. For example, human placental lactogen, acting on maternal prolactin receptors, increases maternal insulin resistance, thereby diverting glucose to the fetus but also, occasionally, leading to gestational diabetes (Haig, 1993).

8.6. CHANGING ENVIRONMENT: COEVOLUTIONARY PROCESSES

We live in a changing environment. Genetic evolution is slower than environmental change, and so is always playing 'catch up' to a changing environment. Our environment comprises an abiotic environment, a non-human biotic environment, and an environment created by humans and their products. Because of our ability to create or construct our environments, the abiotic environment has relatively little effect on human health—fortunately, relatively few people suffer from frostbite or dehydration. Nonetheless, skin cancer and rickets may be thought of as diseases resulting from an excess or a deficiency of ultraviolet radiation. Despite selection for efficient metabolism, humans require some minimal level of nutrients to develop and function normally; natural selection cannot prevent diseases that result from nutritional deficiencies.

8.6.1. The Biotic Environment: Parasites

Most importantly for human health is that component of the biotic environment that comprises our parasites. Humans are host to countless species of microorganisms that have evolved to use our bodies not simply as sources of nutrition but as environments in which to grow and reproduce. Moreover, as we know from the phenomenon of emerging diseases, many microorganisms that now infect other species are only an ecological or evolutionary step away from infecting humans. The selection of pathogens that can live in and on human beings, and that can be transmitted efficiently between humans, and selection for people who are resistant to these pathogens, results in a process of host-parasite coevolution. Because of their large population sizes, high mutation rates, and short generation times, parasites have the advantage in this coevolutionary process. As long as our bodies provide habitats for organisms of other species, these organisms will evolve to utilize our resources for their own growth and replication.

8.6.2. The Human Environment: Cultural Practices

Cultural beliefs, practices, and artifacts form an increasingly important part of the human environment. These cultural practices may also change more rapidly than genes can adapt. The invention of plumbing, and the practice of fermenting fruits together with the development of lead-containing vessels to store and transport the fermented liquid, has resulted in an increased concentration of lead in the environment. No doubt, there is heritable variation in the sensitivity to lead, and with enough time the human species might evolve to have greater resistance to lead than it currently does; in the meantime, however, too many people suffer from lead poisoning. Diseases such as diabetes and hypertension may well result from a culturally-driven changing human environment, in which the availability of food has increased and the need for physical labor to produce food has decreased.

To the extent that people with specific genotypes preferentially reject or adopt specific cultural practices, there is a process of gene-culture coevolution that is analogous to host-parasite coevolution. The classic example of gene-culture coevolution concerns the coevolution of dairying and of lactase persistence, the ability to metabolize lactose in adult life (Durham, 1991). The domestication of cattle and the development of dairying led to the availability of fresh milk as a potential energy source, which in turn led to selection of individuals who could utilize the lactose in milk as a nutrient after the weaning period. Cultures with a high frequency of the lactase persistence allele produced and consumed fresh milk, while populations that had a low frequency of this genotype either didn't milk cattle or developed methods to ferment milk and lower its lactose content. Because cultural traits evolve and spread more rapidly than do alleles, the availability of fresh milk is now more widespread than the trait of lactase persistence; consumption of fresh milk by people without this trait may lead to the gastrointestinal symptoms of lactose intolerance.

8.7. DISCUSSION

From an evolutionary perspective, diseases may have multiple causes. Thus, cancer may be thought of as a disease of aging, as the result of a conflict between levels of selection, as a disease of chance, and as an environmental disease. Cancer results from somatic mutations that go unrepaired. These mutations are themselves stochastic events, but the probability of a mutation occurring may be increased by environmental mutagens, and the probability that the mutation will go unrepaired may be a manifestation of aging, a consequence of the diversion of resources away from somatic repair and toward reproduction. These mutations lead to a breakdown of the normal subordination of the survival of somatic cells to the survival and reproduction of the organism of which they are part.

Evolutionary biology and medicine have developed as distinct disciplines, with distinct concerns. Evolutionary biology is concerned with ultimate causes of biological phenomena, causes that have operated during the phylogenetic history of a species; these are the causes that have led to the variety and diversity in the natural world. In contrast, medicine focuses on proximate causes of disease, causes that operate during the lifetime of an individual, because these are the causal pathways in which medicine can intervene. An evolutionary nosology, a nosology of ultimate causes, complements the traditional medical classification of disease. Classification of a disease as an infectious disease, for example, may suggest that it be treated with antibiotics. On the other hand, understanding infectious diseases as the outcome of host-parasite coevolution not only explains why these diseases persist, but also helps to explain their severity and natural history. The natural histories of infectious diseases depend on the details of the interactions between the parasites and their hosts. Parasites undergo selection both for replication within hosts and for transmission between them; the effects of parasites on their hosts are byproducts of selection for these other traits. In general, pathogens that are transmitted most efficiently from healthy hosts evolve to be benign. On the other hand, pathogens that are transmitted most efficiently from sick or debilitated hosts will evolve to make their hosts sick. Thus, diseases that are spread by direct contact tend to be benign, while diseases that are transmitted via insect vectors are often virulent (Ewald, 1994). Moreover, since pathogens evolve to grow in and to be transmitted between the most abundant genotypes in the host population, they cause frequency-dependent selection of rare host genotypes and therefore promote genetic diversity in their host population (Wills, 1996). This genetic diversity not only provides a population-level defense against the spread of pathogens, but also helps to explain variations in the severity of infectious diseases. Finally, an evolutionary understanding of infectious diseases may not only suggest appropriate regimens of antibiotic usage, but may even lead to strategies for amelioration of the disease (Ewald, 1994).

More broadly, evolutionary considerations remind us of the significance and meaning of human variation, and caution us against the growing practice of labeling variation as pathology, of confusing the normal distribution with normality (Davis and Bradley, 1996). Lastly, if nothing else, an evolutionary nosology helps to clarify the reasons why disease persists, and why disease will always be part of the human condition.

ACKNOWLEDGEMENTS

An earlier version of this paper was presented at a conference, 'Philosophical Issues in the Biomedical Sciences,' sponsored by the Department of Philosophy, University of Alabama at Birmingham, Birmingham AL, May 13–15, 2004, and has been published (Perlman, 2005).

University of Chicago, Chicago, Illinois, USA

REFERENCES

Barker DJP (ed) (1992) Fetal and infant origins of adult disease. British Medical Journal, London
Boorse C (1977) Health as a theoretical concept. **Phil Sci** 44:541–573
Buss LW (1987) The evolution of individuality. Princeton University Press, Princeton
Charlesworth B (1994) Evolution in age-structured populations, 2nd edn. Cambridge University Press, Cambridge
Coall DA, Chisholm JS (2003) Evolutionary perspectives on pregnancy: maternal age at menarche and infant birth weight. Soc Sci Med 57:1771–1781
Cunningham A (1992) Transforming plague: the laboratory and the identity of infectious disease. In: Cunningham A, Williams P (eds) The laboratory revolution in medicine, Cambridge University Press, Cambridge, pp 209–244
Darwin C (1859) On the origin of species. John Murray, London; Repr. Harvard University Press, Cambridge, MA (1964)
Davis PV, Bradley JG (1996) The meaning of normal. Perspect Biol Med 40:68–77
Distante S, Robinson KJH, Graham-Campbell J, Arnaiz-Villena A, Brissot P, Worwood W (2004) The origin and spread of the *HFE*-C282Y haemochromatosis mutation. Hum Genet 115:269–279
Doolitte WF, Sapienza C (1980) Selfish genes, the phenotype paradigm and genome evolution. Nature 284:601–603
Durham WH (1991) Coevolution: genes, culture, and human diversity. Stanford University Press, Stanford
Ewald PW (1994) Evolution of infectious disease. Oxford University Press, New York
Faber KH (1930) Nosography: the evolution of clinical medicine in modern times, 2nd edn. Hoeber, New York
Greaves M. (2002) Cancer causation: the darwinian downside of past success? Lancet oncol 3:244–251
Haig D (1993) Genetic conflicts in human pregnancy. Q Rev Biol 68:495–532
Hales CN, Barker DJP (2001) The thrifty phenotype hypothesis. Br Med Bull 60:5–20
Hesslow G (1993) Do we need a concept of disease? Theor Med 14:1–14
Hill K, Kaplan H (1999) Life history traits in humans: theory and empirical studies. Annu Rev Anthropol 28:397–430

Hull DL (1980) Individuality and evolution. Annual review of ecology and systematics 11:311–332

Kirkwood T (1999) Time of our lives: the science of human ageing. Oxford University Press, New York

Kovács J (1998) The concept of health and disease. Med Health Care Philos 1:31–39

Marmot MG (2003) Understanding social inequalities in health Perspect Biol Med 46:S9–S23

Mayr E (1964) Introduction. In: Darwin C, On the origin of species Repr. Harvard University. Press, Cambridge, pp vii–xxviiMA

Nesse RM (2001) On the difficult of defining disease: a darwinian perspective. Med Health Care Philos 4:37–46

Nesse RM, Williams GC (1994) Why we get sick: the new science of darwinian medicine. Time Books, New York

Online mendelian inheritance in man (OMIM) (2006) Available: http://www.ncbi.nlm.nih.gov/Omim/

Orgel LE, Crick FHC (1980) Selfish DNA: the ultimate parasite. Nature 284:604–607

Perlman RL (2005) Why disease persists: an evolutionary nosology. Med Health Care Philos 8:343–350

Rees J (2002) Complex disease and the new clinical sciences. Science 296:698–701

Sterelny K, Griffiths PE (1999) Sex and death: an introduction to philosophy of biology. University of Chicago Press, Chicago

Weatherall DJ (2001) Phenotype-genotype relationships in monogenic disease: lessons from the thalassaemias. Nat Rev Genet 2:245–255

Wilkinson RG (1997) Unhealthy societies: the afflictions of inequality. Routledge, New York

Williams GC (1957) Pleiotropy, natural selection, and the evolution of senescence. Evolution 11:398–411

Wills C (1996) Yellow fever, black goddess: the coevolution of peoples and plagues. Addison-Wesley, Reading, MA

9. CREATING MENTAL ILLNESS IN NON-DISORDERED COMMUNITY POPULATIONS

9.1. INTRODUCTION

Mental illness is usually viewed as a quality of disordered individuals. Almost all researchers about this topic study the nature, causes, and treatments of psychopathological symptoms. Likewise, commonsense conceptions of mental illness among laypersons also involve problems of specific disordered individuals. Yet, mental illness is also a cultural category referring to social definitions of the phenomenon. From this perspective, the object of study is the cultural definitions and rules that define mental illness, not the individuals who manifest the behaviors to which the rules apply (Horwitz, 1999, pp. 57–78). Social constructionist studies examine the social use of concepts and the way that different groups exploit the ambiguities of concepts such as mental illness. To this end, they study the social forces behind the emergence, maintenance, and change of definitions of mental illness and the ways in which these definitions further the interests of particular groups. Concepts of mental illness, from the constructionist perspective, are aspects of programs of social action not of disturbed individuals (Hacking, 1999). The social constructionist view does not contradict the traditional realist view but examines different aspects of mental illness.

This paper examines one particular feature of current concepts of mental illness: the social construction of how many persons in community populations suffer from these disorders. Mental illnesses are currently viewed as widespread, even rampant, in contemporary American society. The major surveys of the amount of mental illness in the population indicate that about 50 million people suffer from mental disorders each year. These surveys show that about nearly a third of adults suffer from the most common psychiatric illnesses over a twelve-month period and nearly half warrant diagnoses of these illnesses over their lifetime (USDHHS, 1999). Surveys of some groups, such as adolescents and the elderly, uncover even higher rates, often approaching nearly fifty percent over a one-year period (Lavretsky and Kumar, 2002, pp. 239–255; Roberts et al., 1990).

These figures about the prevalence of mental illness are usually taken-for-granted and are routinely cited in scientific studies, media reports, advocacy documents, and pharmaceutical advertisements. They are used to support arguments that mental disorder is a public health problem of vast proportions, that few people with these conditions seek appropriate professional treatment, that untreated disorders create vast economic costs, and that more people need to take medications or seek psychotherapy to overcome their suffering (Greenberg et al., 1993, pp. 405–418; Hirschfield et al., 1997, pp. 333–340; USDHHS, 1999). This paper, in contrast,

123

H. Kincaid and J. McKitrick (eds.), Establishing Medical Reality, 123–135.
© 2007 *Springer.*

argues that the extraordinarily large number of people who suffer from putative mental disorders is largely a product of a particular methodology that inevitably overstates the number of people in the community who have some untreated mental illness. This methodology relies on the presence of symptoms alone without regard for the context in which symptoms develop. It does not allow for the possibility that symptoms could arise for any reason other than the presence of an underlying mental illness. This procedure inevitably results in overstatements of the amount of mental illness because it treats both symptoms that are non-disordered features of human experience and those that are signs of pathology as signs of disorders.

The overstated prevalence estimates that result from community studies are perpetuated because they serve the interests of a number of interest groups including the National Institute of Mental Health, patient advocacy groups, mental health researchers and clinicians, and pharmaceutical companies. Social, political, economic, and cultural factors, not the actual presence of so many mentally disturbed people in the population, account for the acceptance and promotion of high prevalence rates of mental illness. This paper uses the three examples of depression, sexual dysfunction, and social anxiety disorder to illustrate the social construction of widespread mental illness.

Prevalence estimates derive from studies of the amount of mental illness in untreated community populations, not from the number of people who actually seek and enter mental health treatment. The goal of these studies is to determine how many people in the community who do not seek professional treatment have psychological conditions that would meet diagnostic criteria for a mental disorder. To this end, these surveys take the definitions of various psychiatric disorders found in the *Diagnostic and Statistical Manual* (DSM) of the American Psychiatric Association, transform them into standardized questions, and ask randomly selected samples of the population whether or not they have experienced them. A computer program then applies an algorithm to determine if respondents meet the diagnostic criteria.

Epidemiological surveys of prevalence rates that can be generalized to the population must collect thousands of cases. Therefore, they must use many interviewers who ask the same questions. To this end they develop closed-format questions and answers that trained lay interviewers can administer to gather information about symptoms. For prevalence estimates to be accurate, different interviewers must ask these questions in exactly the same way: 'The interviewer reads specific questions and follows positive responses with additional prescribed questions. Each step in the sequence of identifying a psychiatric symptom is fully specified and does not depend upon the judgment of the interviewers' (Leaf et al., 1991, pp. 11–32). This standardization is necessary because only minor variations in question wording, interviewer probes, or instructions can lead to major differences in results. These responses to standardized questions are assumed to provide comparable measures to responses that patients make to clinical interviews.

The rigid standardization of structured interviews improves the consistency of symptom assessment across interviewers and research sites and the consequent reliability of diagnostic decisions (Wittchen, 1994, pp. 57–84). The standardized methods of community surveys are also far cheaper and more efficient than using clinicians to interview thousands of community members. The resulting diagnoses of particular disorders, however reliable they may be, nevertheless do not provide accurate measures of mental illness in community populations.

The core assumption in community studies is that a structured diagnostic interview allow researchers '... to obtain psychiatric diagnoses comparable to those a psychiatric would obtain' (Robins and Regier, 1991). This is because the questions used in community surveys almost precisely match the symptom criteria to make diagnoses found in the DSM. In fact, however, the production of rates of mental disorder in community populations differs in fundamental respects from the study of people who seek clinical mental health treatment. People seeking help from clinicians are highly self-selected and inherently use contextual information about whether or not their conditions exceed ordinary and transitory responses to stressors. A study of people who sought help from psychiatrists illustrates how only symptoms attributed to internal psychological problems and not to stressful situations are brought to professional attention:

> ... once it becomes undeniable that something is really wrong, that one's difficulties are too extreme to be pushed aside as either temporary or reasonable, efforts begin in earnest to solve the problem. Now choices to relieve pain are made with a conscious and urgent deliberation. The shift in thinking often occurs when the presumed cause of pain is removed, but the difficulty persists. Tenure is received, you finally get out of an oppressive home environment, a destructive relationship is finally ended, and so on, but the depression persists. Such events destroy theories about the immediate situational sources of depression and force the unwelcome interpretation that the problem might be permanent and have an internal locus. One has to consider that it might be a problem of the self rather than the situation (Karp, 1996, pp. 112–113).

These interpretative processes lead people to seek professional help only after they make self-definitions that their symptoms arise from some internal problem that is highly distressing or disabling.

Not only patients but also clinicians provide contextual screens for diagnoses of mental illness among treated populations (Wakefield, 1999, pp. 29–57). The primary goal of clinicians is to diagnose and treat a particular individual, not to insure that their diagnosis and treatment conforms to what other clinicians might be doing. Therefore, their purposes differ in basic respects from the goal of standardization that drives community studies. While clinical interviews, like epidemiological surveys, rely on self-reported symptoms, they have built-in probes that assess the circumstances under which symptoms developed. In addition, clinicians can draw on their experience and use their discretion to decide whether symptoms are expressions of an underlying disorder or the result of stressful situations. They thus represent a second layer of judgment about whether a particular person has a mental disorder and, if so, what type of disorder they might have.

A clinical diagnosis thus only arises after people seeking clinical treatment have defined themselves as having a problem and that their problem warrants professional help. Furthermore, the presence of a diagnosis implies that clinicians have used their judgments that symptoms are not simply ordinary responses to life problems among non-disordered people. Considerations of the context within which symptoms develop are built into the diagnostic process for clinical populations.

In contrast to the contextual decisions that drive diagnoses in clinical practice, the diagnostic process in community studies precludes any consideration of the context in which symptoms develop. Interviewers are forbidden from making any judgments about the validity of respondent answers and can neither exercise clinical discretion nor use flexible probes about responses. Even if the respondent seems to have misunderstood the standardized question, the interviewer is instructed to repeat the question verbatim (Brugha et al., 1999, pp. 1013–1020). The absence of interviewer probes can produce seriously misleading results. For example, a person asked 'Have you ever had a period of two weeks or more when you had trouble sleeping' might recall a period when ongoing construction outside their bedroom interrupted their sleep. Another, when asked 'How often do you hear voices that other people say they cannot hear?' might recall his hard-of-hearing wife's inability to hear their neighbors' late night arguments (Link, 2002, pp. 247–253). Faced with such situations, respondents can disregard the literal meaning of the question, self-censor their response, and not report the 'symptom.' Alternatively, they can provide an answer that is literally true so that their troubled sleep or hearing of voices will be counted as a symptom of a mental illness. Interviewers cannot, however, respond to a question from respondents about whether this is the kind of situation that they are trying to collect information about. They must record all positive responses, regardless of their context, as symptoms of mental disorders.

The a-contextual measures of community surveys preclude them from treating 'symptoms' as responses that any normal, non-disordered person would make to given situations; all symptoms, regardless of the context in which they emerge and persist, *must* be seen as indicators of mental disorders. Treating both non-disordered, as well as dysfunctional, responses as signs of pathology, in turn, inherently inflates rates of putative pathological conditions (Wakefield, 1999). For example, the estimates of mental illness in one major epidemiological study are nearly twice as high as those produced when psychiatrists interview the same people (Anthony et al., 1985, pp. 667–675). Community studies of rates of depression, sexual dysfunction, and social anxiety disorder illustrate how the methods that these studies use are responsible for the large prevalence rates that these studies uncover.

9.2. DEPRESSION

The most widely used estimates of the prevalence of depression in the United States stem from the National Comorbidity Survey (NCS). Its findings and those from a similar study, the Epidemiologic Catchment Area (ECA) study are the basis for the estimates regarding the frequency of mental disorder that are now widely cited in the scientific, policy, and popular literatures. The NCS uses two steps to obtain diagnoses of depression based on DSM-IIIr criteria (Blazer et al., 1994, pp. 979–986). The first is that respondents must affirm at least one stem question that appears at the beginning of the interview. These questions ask: 'In your lifetime, have you ever had two weeks or more when nearly every day you felt sad, blue, or depressed?'; 'Have you ever had two weeks or more when nearly every day you felt down in the dumps, low, or gloomy?'; 'Has there ever been two weeks or more when you lost interest in most things like work, hobbies, or things you usually liked to do for?'; and 'Have you ever had two weeks or more during which you felt sad, blue, depressed or where you lost all interest and pleasure in things that you usually cared about or enjoyed?' Given the broad nature of these questions and the fact that they have no exclusion criteria for the circumstances in which these moods arose, it is not surprising that 56% of the population reports at least one 'yes' response to them (Blazer et al., 1994). Later in the interview, this group is asked questions about symptoms that derive from the DSM criteria for Major Depression. To receive a diagnosis of depression, community members must report having depressed mood or inability to feel pleasure and four additional symptoms such as loss of appetite, sleep difficulties, fatigue, or inability to concentrate on ordinary activities. A computer program then determines if respondents meet the criteria for a diagnosis of depression.

The NCS estimates that about 5% of subjects have a current (30-day) episode of major depression, about 10% had this condition in the past year, about 17% at some point in their life, and about 24% reported enough symptoms for a lifetime diagnosis of either major depression or dysthymia (Blazer et al., 1994; Kessler et al., 1994, pp. 8–19). It also finds that relatively few people who are diagnosed with these conditions have sought professional help for them. About a third of persons with diagnosis of Major Depressive Disorder had sought any kind of professional treatment for their condition and far fewer sought help from mental health professionals (Kessler et al., 1994).[1]

Are the many cases of Major Depression uncovered in community studies equivalent to treated clinical cases? Unlike the situation in clinical practice, current community surveys cannot separate symptoms that are expectable responses to particular situations from those that are disorders. In contrast to clinical settings where the judgments of both lay persons and clinicians distinguish ordinary sadness from depressive disorders, symptom-based diagnoses in community studies consider all persons who report enough symptoms as having the mental disorder of depression. A respondent might recall symptoms such as depressed mood, insomnia, loss of appetite, or diminished pleasure in usual activities that lasted for longer than two weeks after the breakup of a romantic relationship, the diagnosis of a serious illness in an intimate, or the unexpected loss of a job. Although these symptoms might have dissipated as soon as a new relationship developed, the intimate recovered, or another job was found, this individual would join the 20 million people who suffer from the presumed disorder of depression each year. For example, in the ECA study the most common symptoms among those reporting symptoms are 'trouble falling asleep, staying asleep, or waking up early' (33.7%), being 'tired out all the time' (22.8%), and 'thought a lot about death' (22.6%) (Judd et al., 1994, pp. 18–28). College students during exam periods, people who must work overtime, those worrying about an important upcoming event, or respondents taking a survey around the time of the death of a famous person would all naturally experience these symptoms. Symptoms that neither respondents nor clinicians would consider being reasons for entering treatment nevertheless are treated as signs of disorder in community surveys. Moreover, the duration criteria only require that the symptom last for a two-week period, insuring that many transient and self-correcting symptoms are counted as disordered. Standardization produces not only more reporting of depressive symptoms but also of non-disordered responses, without distinguishing the two.

Current epidemiological studies of depression treat all positive responses to standardized questions as signs of disorder, no matter what their cause or context may be. In the absence of knowledge about the context in which symptoms arose and persisted, there is no way to know if any symptom is an indicator of a disorder or a natural response to a stressful situation. Community studies should accurately capture people who actually have depressive disorders and so should not have many false negatives – people who are truly depressed but are not found to be depressed in surveys. However, the inflexibility of symptom ascertainment greatly increases the chances of false positive diagnoses – counting people who do not have a depressive disorder as depressed – because they count all symptoms, including those that stem from adaptive responses to situational stressors, as indicators of disorders. Because the kinds of symptoms that presumably indicate depressive disorders are often common products of ordinary stressors the number people who do not have mental disorders but are diagnosed as depressed might even exceed the number of people who are accurately classified as depressed (Anthony et al., 1985; Wakefield, 1999). Diagnostically-oriented community studies did not so much uncover high rates of depressive disorders as they demonstrated that the natural results of stressful social experiences could be distressing enough to meet symptom criteria for a disorder.

9.3. SEXUAL DYSFUNCTION

A study of sexual dysfunction based on the largest and most representative community study of sexual behavior ever conducted in the United States also illustrates how the use of a-contextual, symptom based logic in survey research inflates rates of presumed mental disorders (Laumann et al., 1999, pp. 537–544). The study was published in the prestigious *Journal of the American Medical Association (JAMA)* and featured on the front page of the *New York Times*, and its results continue to be widely noted and disseminated. Comparable to diagnoses of depression, it measures sexual dysfunction with seven symptoms that correspond to DSM criteria including a lack of interest in sex, anxiety about sexual performance, arousal difficulties, the inability to have an orgasm, difficulty in obtaining erection, or finding sex painful or not pleasurable. It finds that 43% of women and

31% of men suffered from sexual dysfunctions over the past 12 months. These huge numbers result from a failure to use a valid definition of mental disorder that distinguishes internal dysfunctions from expectable results of social stressors or of diminished, but normal, interest in having sex.

Symptoms such as the failure to obtain an erection, to find sex pleasurable, or to be orgasmic can sometimes indicate harmful internal dysfunctions and, hence, can be signs of a mental disorder. They would be dysfunctions when people are unable to engage in sexual activity despite their desire to have sex with a particular partner. They could also, however, arise from situations that are not related to any pathological individual condition. For example, people who have boring, inexperienced, or inept sexual partners might naturally report experiencing some of these symptoms. Others, who are in unsatisfactory or abusive relationships, would naturally have sexual difficulties with undesirable partners. Having *partners* with sexual dysfunctions could account for still other symptoms. Even some well-adjusted couples that no longer have a strong interest in pursuing sexual relationships could report symptoms of supposed dysfunction.

The primary cause of putative sexual dysfunctions that this study finds confirms the suspicion that symptom-based measures conflate non-disordered with disordered conditions. The best predictor of 'sexual dysfunction' is low satisfaction with one's sexual partner: sexual problems stem from difficult interpersonal relationships. If people were to regain normal sexual functioning when they change sexual partners or when their relationship with their current partner improves, their current symptoms would not indicate an internal dysfunction but instead would reflect a problematic social relationship. The assumption that the presence of symptoms, regardless of what factors account for them, represents an individual dysfunction is unwarranted. Symptom-based definitions of sexual dysfunction hopelessly entangle people whose symptoms actually do stem from internal dysfunctions with those whose symptoms are not the result of any psychological or physiological dysfunction.

The problem with the assumption that all symptoms indicate mental disorder regardless of the meaning or context of these symptoms is not only the resulting inflated rates of disorder but also the type of solution that appears to follow. The authors of the *JAMA* study conclude that sexual dysfunction is a 'public health' problem that calls for increased provision of medical therapies, especially medications. They suggest that people with bad or boring interpersonal relationships, as well as those with dysfunctions that preclude them from having desired sexual relationships, should remedy their condition through taking drugs that increase their sexual stimulation. Their findings, however, suggest that people would be better advised either to change their relationships with their current partners or to find different partners instead of seeking medication from their physicians. Alternatively, they might be perfectly satisfied with relationships that don't involve much sexual activity. Community studies can easily generate very high, but misleading, prevalence rates: there is no valid reason to consider nearly half of women and one third of men as suffering from the disorder of sexual dysfunction.

9.4. SOCIAL ANXIETY DISORDER

Social anxiety disorders, also called social phobias, are conditions that feature marked and persistent fears of social or performance situations in which embarrassment may occur (APA, 1994, p. 411). The NCS found that social phobias are among the most common mental disorders, exceeded in number only by depression and alcohol problems. Pervasive television and print advertisements encourage sufferers to seek medical help for them. Widespread public service announcements likewise urge the many millions of people afflicted with social anxiety disorders to recognize that they have real disorders. An annual National Anxiety Disorders Screening Day has been established to enhance awareness and professional help-seeking for this disorder.

Social anxiety did not exist as an officially recognized disorder until 1980. The DSM-I and DSM-II did not mention them and they first entered the psychiatric literature only in the late 1960s (Healy, 1997). When they first appeared in the DSM-III in 1980, the manual noted that: 'The disorder is apparently relatively rare (APA, 1980, p. 228).' The Epidemiological Catchment Area study provided the first prevalence estimates for social phobias of about 2.75% in community populations during the early 1980s (Eaton et al., 1991). This disorder did not emerge as a widely prevalent mental illness until the results of the National Comorbidity Survey were published in the early 1990s. The NCS estimated their lifetime prevalence at 13.3%, one out of every eight people in the population and about five times higher than the rates in the prior ECA study (Magee et al., 1996).

What led to the near quintupling in the prevalence of social phobias over this ten-year period? Survey questions changed the criteria for a diagnosis of social phobia from requiring a compelling desire to avoid exposure to social or performance situations to needing only marked distress in these situations. In the NCS, people received diagnoses of social phobias when they reported an unreasonable fear that leads them to avoid or to feel extremely uncomfortable while doing at least one of the following: public speaking, using the toilet when away from home, eating or drinking in public, feeling foolish when speaking to others, writing while someone watches, or talking in front of small groups of people. The most prevalent responses leading to a diagnosis affirm the question of having an unreasonably strong fear of speaking in public (McHugh, 1999, pp. 32–38).

It is not surprising that one of eight persons, when reflecting over a lifetime, would recollect feeling intensely nervous before speaking at an important meeting, having nothing to say on a date, or fearfully approaching an oral in-class presentation. Most people, however, rarely have any activities that require them to speak very often to audiences larger than two or three family, friends, or colleagues. Their responses to questions that ask: 'Have there ever been times when ...' would naturally refer to unusual situations when they had to toast their sister at a wedding, honor their parents at an anniversary celebration, or present a report in a school classroom. Such questions and subsequent diagnoses will certainly generate huge prevalence estimates, but these estimates will not be valid estimates of social anxiety as mental disorders. The symptom-based logic of community studies has created these disorders as commonly occurring conditions.

Salespeople, teachers, or executives whose jobs require them to speak before large audiences would be well advised to seek professional help when they experience the symptoms of social anxiety. Others might justifiably question if they have a mental disorder or, alternatively, ordinary experiences of discomfort that do not seriously disrupt their normal functioning. In the past, people who reported these symptoms rarely sought mental health services: persons with social phobias have a lower rate of help-seeking from mental health professionals than any other disorder except for substance abuse (Katz et al., 1997, pp. 1136–1143). Now, however, many groups have an interest in promoting the pervasiveness of social anxiety disorders and the need to seek professional treatment for them.

9.5. WHY ARE HIGH PREVALENCE RATES PERPETUATED?

It would be possible for community surveys to separate non-disordered responses to problematic situations from mental disorders far more adequately than they do at present. They could build in questions that ask about the context in which symptoms develop and persist. Such questions would ask respondents if symptoms of, for example, depression emerged during periods of intense stress and disappeared as soon as these periods of crises were over. Likewise, surveys could separate the inability to achieve erections from the lack of interest in a particular sexual partner by incorporating questions about the context in which the lack of sexual performance occurs. Contextual probes could also separate persons who are shy or nervous from those who are so fearful that they are unable to engage in ordinary social interactions. Clinical treatment routinely features such probes and they are compatible with basic principles of survey methodology (Brugha et al., 1999). The reason why more contextual exclusion criteria are not built into community surveys has to do not only with the efficiency and practicality of a-contextual, standardized methods but also with the benefits that particular groups get from high prevalence rates of mental illnesses.

Several different, but interlocked, groups promote estimates that emphasize how common mental disorders are in community populations. The National Institute of Mental Health (NIMH) is the major funder of research about the frequency of mental illness and a central sponsor of the notion that mental illnesses are widespread in the population. Its historical development shows why this agency advocates high prevalence rates of particular disorders. During its initial decades the NIMH promoted an expansive agenda of community mental health, which encompassed broad policy objectives (Grob, 1991). In the 1960s the NIMH sponsored projects that attempted to alleviate poverty, combat juvenile delinquency, and promote social change. Political changes in the 1970s, however, forced the NIMH to change its focus from social and economic problems to specific disease conditions (Kirk, 1999). Studying and treating specific diseases was far more politically palatable than addressing controversial social problems.

The epidemiological studies in the 1980s and 1990s were part of the NIMH's efforts to show how presumed disease conditions did not just afflict a small group of persons but were widespread, yet untreated, problems in the population. Conflating mental disorders and problems in living, while calling both *mental illnesses*, insulated the agency from political pressures, expanded its mandate, enhanced the importance of the problem it addressed, and protected its budget (Horwitz, 2002). Political support is far more likely to accrue to an agency that is devoted to preventing and curing widespread disease conditions than to one that confronts controversial social problems.

Pharmaceutical companies also have a major interest in showing that mental diseases are widespread in the population. Although there is little evidence that these companies influenced the methods of epidemiological surveys that produce high prevalence rates, they have capitalized on the findings of these surveys because they create a broader market of diseases for their products to treat. Pharmaceutical advertisements routinely feature the alleged numbers of people who suffer from particular mental disorders, sending the message that potential consumers are not unique but share their problems with millions of others. These estimates provide a valuable rhetorical tool for expanding the market for the various medications these companies produce. The explosive growth in sales of antidepressant medications is testimony to the effectiveness of this appeal (Healy, 1997; Horwitz, 2002).

Social anxiety disorders illustrate this process. As noted above, by slightly changing the wording of questions that establish the criteria for this disorder, the NCS presumably established that one of every eight people suffered from this disorder. This created a fertile new market for the marketing of SSRI anti-depressant medications, which were becoming increasingly popular in the early 1990s. In 1999 the SSRI, Paxil, was approved for the specific treatment of social anxiety, unleashing a barrage of print and television ads. The vast potential market for this medication is less the small number of persons with severe social phobias who are in clinical treatment than the far greater number of persons that community studies have uncovered (Raghunathan, 1999). The company attempts to reach untreated sufferers through broadcasting the prevalence estimates of community studies and showing how common these conditions are. One television advertisement for Paxil, for example, features an attractive woman who is extremely nervous before speaking at a gathering of her extended family that might be an anniversary, wedding, or birthday. Such portrayals attempt to convince people both that their discomfort is a mental disorder and that it is an extremely common condition. The use of high frequency estimates of conditions such as social anxiety disorders can, paradoxically, serve to create the very phenomenon they claim to be describing.

The presumably enormous number of sufferers from sexual dysfunction disorders also served to create a large market for pharmaceutical products. The *JAMA* study of sexual dysfunction was published shortly after a new drug, Viagra, came on the market, with sales that exceeded $1 billion in its first year. Two of the authors of the article are consultants to Pfizer, the maker of Viagra (Rosen, 1999). When Viagra was first marketed it was directed toward older men who had problems achieving sexual climax. Robert Dole, the former Republican presidential candidate, was the major initial spokesman for the benefits of Viagra. Shortly thereafter, it became clear that older persons formed only a small portion of the potential market for Viagra because young, sexually capable, men were widely using the drug to enhance their sexual performance (Hitt, 2000, 34ff.). Subsequent advertising was directed at this much larger market, using attractive young people as models and asking 'Not satisfied with your sex life?' Studies that find high prevalence estimates of supposed sexual dysfunctions serve to justify the expansion of the market for Viagra (and now also its major competitors, Levitra and Cyalis) well beyond persons with erectile dysfunctions to any man who wants enhanced sexual performance.[2] Calling people with problems in their interpersonal relationships 'sexually dysfunctional' may help the business of the pharmaceutical company that sponsored this research, but it fundamentally mischaracterizes the nature of most of the problems that the study uncovers.

Family advocacy groups, which became influential during the 1980s, are another interest group that promotes the idea that mental illnesses such as depression and anxiety are widespread in the population. They believe that demonstrating the ubiquity of mental illness in the community will aid their efforts to destigmatize this condition and obtain more resources for its treatment. Groups such as the National Alliance for the Mentally Ill and the Anxiety Disorders Association of America embrace claims about the widespread nature of mental disorders because the many millions of people that community surveys identify can be equated with the far smaller number of people with truly serious mental disorders. This equation presumably reduces the social distance between the mentally disordered and others and lowers the stigma accorded mental illness.

These groups promote high prevalence rates because they think that if they can convince politicians that mental illnesses are widespread, rather than uncommon, they can increase the amount of funding for mental health services. It is just as likely, however, that their efforts to get more treatment for currently untreated cases will shift resources away from people who truly need professional mental health services toward those who might be distressed but are not disordered (Mechanic, 2003, pp. 8–20). Erasing the distinction between non-disordered and dysfunctional conditions and calling both *mental disorders* can have the counterproductive result of harming the truly disabled.

9.6. CONCLUSION

Estimates of the frequency of mental disorders emerge, persist, and change because of the agendas of particular social groups. While these groups have separate agendas, government scientists and bureaucrats, advocacy groups, and pharmaceutical companies commonly join together in efforts to promote the ubiquitous nature of mental illness (see especially Hirschfield et al., 1997). Because these estimates bring benefits to particular interest groups does not mean that they are *only* social constructions. Many people actually have major depression, sexual dysfunction, or social anxiety disorder. However, the ways in which prevalence estimates of these disorders are constructed and used reflect how different interest groups are able to exploit the inherent ambiguities of the concept of mental illness to make disorders seem far more widespread than they actually are. The apparent ubiquity of mental illnesses stems from the methods that produce them, not from the actual frequency of these disorders. Mental disorders might be properties of particular individuals but a variety of social groups have reasons to construct them in ways that suit their programs, interests, and goals. Ultimately, the efforts of these groups can actually produce more of these disorders as people change their self-conceptions to conform to the images of ubiquitous mental disorders that are widely broadcast throughout the culture.

Institute for Health, Health Care Policy, and Aging Research and Department of Sociology, Rutgers University, New Jersey, USA

NOTES

[1] The second wave of the NCS conducted in the early 2000s found that rates of professional help-seeking for depression had increased considerably (Kessler et al., 2003).
[2] Pharmaceutical companies as yet have been unable to develop a product that would capitalize on the enormous market of women who presumably have sexual dysfunctions.

REFERENCES

American Psychiatric Association, APA (1980) Diagnostic and statistical manual of mental disorders, 3rd edn. American Psychiatric Association, Washington, DC
American Psychiatric Association, APA (1994) Diagnostic and statistical manual of mental disorders, 4th edn. American Psychiatric Association, Washington, DC
Anthony JC, Folstein MF, Romanoski AJ et al (1985) Comparison of lay diagnostic interview schedule and a standardized psychiatric diagnosis. Arch Gen Psychiatry 42:667–675
Blazer DG, Kessler RC, McGonagle KA, and Swartz MS (1994) The prevalence and distribution of major depression in a national community sample: the national comorbidity survey. Am J Psychiatry 151:979–986
Brugha TS, Bebbington PE, and Jenkins R (1999) A difference that matters: comparisons of structured and semi-structured psychiatric diagnostic interviews in the general population. Psychol Med 29:1013–1020
Eaton WW, Dryman A, and Weissman MM (1991) Panic and Phobia. In: Regier LRD (ed) Psychiatric disorders in America. Free Press, New York, pp 155–179
Greenberg PE, Stiglin LE, Finkelstein, SN, and Berndt ER (1993) The economic burden of depression in 1990. J Clin Psychiatry 54:405–418
Grob G (1991) From asylum to community: mental health policy in modern America. Princeton University Press, Princeton
Hacking I (1999) The social construction of what? Harvard University Press, Cambridge
Healy D (1997) The anti-depressant era. Harvard University Press, Cambridge
Hirschfeld RM, Keller MB, Panico S et al (1997) The national depressive and manic-depressive association consensus statement on the undertreatment of depression. JAMA 277:333–340
Hitt J (2000) The second sexual revolution. New York Times Magazine 20 February, 34ff

Horwitz AV (1999) The sociological study of mental illness: a critique and sythesis of four perspectives. In: Aneshensel CS, Phelan JC (eds) Handbook of the sociology of mental health. Plenum, New York, pp 57–78

Horwitz AV (2002) Creating mental illness. University of Chicago Press, Chicago

Judd LL, Rapaport MH, Paulus MP, and Brown JL (1994) Subsyndromal symptomatic depression: A new mood disorder? J Clin Psychiatry 55:18–28

Karp DA (1996) *Speaking of sadness.* Oxford University Press, New York

Katz SJ, Kessler RC, Frank RG, Leaf P, Lin E, and Edlund M (1997) Utilization of mental health services in the United States and Ontario: the impact of mental health morbidity and perceived need for care. Am J Public Health 87:1136–1143

Kessler RC, McGonagle KA, Zhao S et al (1994) Lifetime and 12-month prevalence of DSM-III-R psychiatric disorders in the United States: results from the National Comorbidity Survey. Arch Gen Psychiatry 51:8–19

Kessler RC, Berglund P, Demier O et al (2003) The epidemiology of major depressive disorder: results from the National Comorbidity Survey Replication (NCS-R). JAMA 289:3095–3105

Kirk SA (1999) Instituting madness: the evolution of a federal agency. In: Aneschensel C, Phelan J (eds) *Handbook of the sociology of mental health.* Plenum, New York, pp 539–562

Laumann EO, Paik A, and Rosen RC (1999) Sexual dysfunction in the United States: prevalence and predictors. JAMA 281:537–544

Lavretsky H, Kumar A (2002) Clinically significant non-major depression: old concepts, new insights. Am J Geriatr Psychiatry 20:239–255

Leaf, PJ, Myers JK, and McEvoy LT (1991) Procedures used in the epidemiologic catchment area study. In: Regier LRD (ed) Psychiatric disorders in America, The Free Press, New York, pp 11–32

Link B G (2002) The challenge of the dependent variable. J Health Soc Behav 43:247–253

Magee WJ, Eaton WW, Witchen HJ, McGonagle KA, and Kessler RC (1996) Agoraphobia, simple phobia, and social phobia in the National Comorbidity Survey. Arch Gen Psychiatry 53:159–168

McHugh PR (1999) How psychiatry lost its way. Commentary: 32–38

Mechanic D (2003) Is the prevalence of mental disorders a good measure of the need for services? Health Aff 22:8–20

Raghunathan A (1999) A bold rush to sell drugs to the shy. New York Times, 18 May, C1

Roberts RE, Andrews JA, Lewinsohn PM, and Hops H (1990) Assessment of depression in adolescents using the center for epidemiological studies depression scale: Psychological Assessment. A Journal of Consulting and Clinical Psychology 2:122–128

Robins L, Regier D (1991) Psychiatric disorders in America: the epidemiological catchment area study. The Free Press, New York

Rosen R (1999) Correction. JAMA 281:1174

U.S. Department of Health and Human Services, USDHHS (1999) Mental health: a report of the surgeon general. National Institute of Mental Health, Rockville, MD

Wakefield JK (1999) The measurement of mental disorder. In: Horwitz AV, Scheid TL (eds) A handbook for the study of mental health: social contexts, theories, and systems, Cambridge University Press, New York, pp 29–57

Wittchen H-U (1994) Reliability and validity studies of the WHO-Composite International Diagnostic Interview (CIDI) :a critical review. J Psychiatr Res 28:57–84

10. GENDER IDENTITY DISORDER

10.1. INTRODUCTION

According to the DSM IV, a person with GID is a male or female that feels a strong identification with the opposite sex and experiences considerable stress because of their actual sex (Task Force on DSM-IV and American Psychiatric Association, 2000). The way GID is characterized by health professionals, patients, and lay people belies certain assumptions about gender that are strongly held, yet nevertheless questionable. The phenomena of transsexuality and sex-reassignment surgery puts into stark relief the following question: 'What does it mean to be male or female?' But while the answer to that question may be informed by contemplation of GID, we should also be aware that the answer to the question 'what does it mean to have GID?' is shaped by our concepts of male and female.

First, I consider the concept of transsexuality, and explain how it forces us to clarify our concepts of sex and gender, and leads to the development of what I will call the 'standard view.' I then explain GID from a mental-health standpoint, question the concept of gender identity, and try to uncover some fundamental assumptions of the standard view. I argue that these assumptions are at odds with the plausible view that gender supervenes on physical, psychological, and/or social properties. I go on to argue, contra the standard view, that gender has no essence. I suggest an anti-essentialist account of gender according to which 'man' and 'woman' are cluster concepts. This undermines the dualistic conception of gender that grounds the standard view. An anti-essentialist view of gender cannot make sense of the concept of 'gender identity' and hence sees so-called 'GID' as primarily conflict between the individual and her society, and only derivatively a conflict between the individual and her body.

10.2. TRANSSEXUALITY AND THE CONCEPTS OF MALE AND FEMALE

Consideration of transsexuality both reveals and challenges assumptions about what it means to be male or female. The expressions 'sex change' and 'sex reassignment surgery' suggest a person goes into surgery as one sex and after the procedure emerges as the other sex. This assumes that the features that are removed are sufficient to be a member of the former sex, and the features that are added are features that are sufficient to be a member the new sex. So, a man can become a woman by first ceasing to be a man, which is accomplished by castration. Breast implants and a vaginoplasty complete the transition. According to this standpoint, 'male' and 'female,' 'man' and 'woman' are defined in terms of primary and secondary sex anatomy.

H. Kincaid and J. McKitrick (eds.), Establishing Medical Reality, 137–148.
© 2007 *Springer.*

However, most people have views that are somewhat more complex. Adam/Linda Parascandola, a biological female, says: 'I knew from a young age that I was male despite all external appearances' (Parascandola, 2001). The documentary 'The Opposite Sex' characterizes a preoperative FtM as 'a man with a very serious birth defect' (2004b). According to MtF Jennifer Diane Reitz 'a transsexual is a mind that is literally, physically, trapped in a body of the opposite sex' (Reitz, 2004). Parascandola articulates the view as follows: 'men and women are the only two types of humans and ... transsexuals have simply had their wires crossed and belong to the sex opposite to the one they were born into' (Parascandola, 2001).

While initially one might think that 'man' and 'woman' are two simple categories, transsexuality forces us to realize that the defining features of these categories are not always found together. We are compelled to refine our concepts and make finer distinctions – distinguishing biological sex from psychological sex, in other words, distinguishing 'sex' from 'gender.' While sex refers to the biological characteristics that mark one as male or female, gender is a collection of behavioral and personality features differentially associated with a particular sex. We might further distinguish sex and gender from 'gender role' – a set of behaviors, relationships, responsibilities and expectations that are more typical of a male or female within a given society (Horvath, 1999). Gender roles typically include sexual behaviors and relationships.

We can now use these distinctions to characterize the 'standard view' of transsexuality. According to Transsexual.org, 'in a nutshell: transsexuality means having the wrong body for the gender one really is.' The preoperative transsexual is in 'a state of conflict between gender and physical sex' (Reitz, 2004). He wants to function in the gender role appropriate to his gender, not his sex. He wants his body to approximate as much as possible the type of body that is appropriate for his gender, and that would allow him to 'pass' in his preferred gender role. Back in 1959, MtF Tamara Rees put it like this:

This surgery does not create a woman where the patient was once a biological male, nor can the patient ever hope to have children. It merely brings the physical appearance of the patient into harmony with the mental pattern (Rees, 1959).

Approximately 40,000 (1/2,500) U.S. males have undergone sex reassignment surgery (Conway, 2002). While approximately 1/3 as many women undergo surgery, it is estimated that many more pass as males without surgical or clinical intervention.

GID is the more general disorder of which transsexuality is the most extreme type. Estimates of the prevalence of GID range from 1 in 30,000 to 1 in 500 (2004a; Conway, 2002). Many patients have milder forms of the disorder, often undiagnosed. Indications of GID include: a strong and persistent cross-gender identification; a stated desire to be the other sex; frequent passing as the other sex; a desire to live or be treated as the other sex; a conviction that s/he has the typical feelings and reactions of the other sex; a persistent discomfort with his/her sex; a sense of inappropriateness of the gender role of his/her own sex; a belief that s/he was born the wrong sex; a preoccupation with changing primary and secondary sex characteristics; and lack of any physical intersex condition. (If a person meeting the above criteria were of ambiguous sex biologically, s/he would not be diagnosed with GID) (Task Force on DSM-IV and American Psychiatric Association, 2000). The standard of care for patients with GID is to offer three stages of therapy: hormones, life experiences in the desired gender role, and sex reassignment surgery, though not all persons with GID want or require all three elements of the therapy (2004a).

10.3. WHAT IS GENDER IDENTITY?

As stated, the GID patient suffers from 'cross-gender identification' or 'strong identification with the opposite sex.' But what does it mean to identify with the opposite sex? In general, to identify with a group is to feel you are similar to members of that group and that you are or should be part of that group. So, to identify with males is to feel that you are similar to men in some important way and that you are or should be a man. When a biologically male has a cross-gender identification, she believes that she is or should be a woman. That is to say, she has a *female gender identity*.

Obviously, the concept of 'gender identity' is central to understanding GID. Kohlberg defines 'gender identity' as 'the ability to discriminate between males and females and to accurately identify one's own sex' (Kohlberg, 1966). If we assume that an accurate identification of one's sex is supposed to be in accord with one's biological sex, then it would seem that GID patients lack this ability to some extent. Hence, an implication of Kohlberg's definition is that GID patients have an inadequate or nonexistent gender identity. But most people think that the GID patient has a gender identity, but one that is atypical for his or her biological sex. Renowned sex and gender psychologist John Money describes gender identity as one's inner sense that one is male or female (Money, 1976). Since persons with GID typically report an early awareness of cross-gender identification, and since psychologists have had little success at changing a patient's gender to fit his or her sex, it is thought that each individual has a gender identity that is formed before birth or early in childhood, and that identity is immutable in later life.

For most women, a female gender identity is possessed in conjunction with a certain cluster of anatomical features (including breasts and ovaries) and social roles (such as care-giver and unpaid domestic laborer). But according to the standard view of GID, one can have a female gender identity even if one is a biological male. And, since someone with a female gender identity can be living one's life as a man, it seems that one can have a female gender identity even if one does not play the social role associated with being female. Note that having a feminine gender identity is not the same thing as merely being effeminate. A man can be effeminate without thinking that he is or should be a woman, and a biological male can display predominately masculine characteristics while claiming to have a female gender identity. Many MtF transsexuals have to learn feminine mannerisms and gestures in order to pass as women. It follows that one can have a female gender identity even if one does not tend to behave as most women do.

This leaves the nature of gender identity somewhat mysterious. Supposedly, the biological female with GID has an inner sense that he is male. To which I ask 'he has an inner sense that he is *what*, exactly?' Does he have an inner sense that he is biologically male? That he has a Y chromosome, testicular tissue, a penis, a preponderance of testosterone in his system, etc.? No. Usually, he is aware that he does *not* have these features. If he had a persistent belief that he did have these features, despite all manner of empirical evidence to the contrary, then GID would be akin to a delusional condition such as 'somatozation disorder' or 'body image distortion,' which is unlikely. The FtM knows he is not biologically male – that's the source of his distress. So, his sense that he is male is not a belief that he is biologically male.

Perhaps he has a sense that he is *psychologically* male. But what does it mean to be psychologically male? Earlier, we equated being psychologically male with having a male gender identity. But if having a male gender identity is having an inner sense that one is psychologically male, we have a tight circle of inter-defined terms, and the nature of 'gender identity' remains mysterious. In saying that a biological female has an inner sense that he is male, what belief are we attributing to him? Perhaps it is the belief that he possesses some ineffable essence of masculinity.

The standard view seems to regard gender identity as the essence of what it means to be a man or a woman. It is supposed that this essence can be possessed in the face of contravening biological, anatomical, psychological, behavioral, and social features. It is further supposed that gender comes in just two types, masculine and feminine, and that a masculine gender is appropriate for a male body and a masculine gender role, and a feminine gender is appropriate for a female body and a feminine gender role. Given that gender role includes sexual behaviors and relationships, the standard view also involves the assumption that someone with a female gender identity will be sexually and romantically interested in males, and that someone with a male gender identity will be likewise interested in females. This would explain the longstanding suspicion that homosexuality is a type of GID (Horvath, 1999). It also explains the fact that attraction to the opposite biological sex is considered a maladjustment for transsexuals (Meyerowitz, 2002). (The comedian Stephen Wright jokes 'I'm a lesbian trapped in a man's body.' Why is this supposed to be funny? Because of a failure to distinguish sexual orientation from gender identity.)

10.4. GENDER SUPERVENES

The assumptions underlying the standard view are at odds with some plausible claims about the nature of gender. One of these claims is that, whatever gender is, it supervenes on biological, psychological, and/or social properties. If you fix the biological, psychological, and social facts, you thereby fix the gender facts. Gender facts are not 'further facts.' If you know an individual's genetic makeup, anatomical features, psychological and behavioral profile, role in a particular society, and the gender norms of that society, you know more than all there is to know about that person's gender in that society.

If a female gender identity is something each human being either does or does not have, what does it supervene on? What is it based on? If, as the standard view claims, a biological male can have a female gender identity, then obviously gender identity cannot supervene on the typical markers of biological sex, such as chromosomes, reproductive organs, and the like. I'll consider four alternative suggestions.

Proponents of the standard view talk as if there is some irreducible fact of the matter as to a person's gender identity. So, one possible view is that gender identity doesn't supervene on anything. Is it a *sui generous* property that does not yield to further analysis. This is a flat denial of the idea that gender supervenes on biological, psychological, and/or social properties. But if gender doesn't supervene, then perfect duplicates could differ with respect to gender. (Interestingly, there have been a few documented cases of homozygous twins raised together, where one twin reported a cross-gender identification and the other did not. However, this is not a case of perfect duplicates differing with respect to gender. Clearly, there was some psychological difference between them, but the cause of that psychological difference is not known (2002).) If gender does not supervene, then gender is disconnected from everything we can know about a person – their physical, psychological, and relational properties – and it is not clear how the concept could be fit for any role in psychological theory or practice.

A related view is that a having female gender identity is not based on anything physical; rather, it is based on having a female soul. This response couples male/female dualism with mind/body dualism. I suspect this 'mind/body/gender' dualism, as I call it, is a deep unarticulated ideology that underlies many of the thoughts that people have on issues of gender. But what makes a soul a female soul, rather than some other gender? Perhaps the answer is as ineffable as souls themselves. In that case, this view is as unsatisfying as the view that gender is *sui generous*, for it leaves gender floating free from a person's knowable characteristics. Perhaps a better answer is that a female soul has certain distinctive tendencies, desires, and beliefs. That would be to say, in short, that what makes a soul a female soul is its psychological properties. If this is the view, then it would be better expressed by saying that gender supervenes on psychological properties, leaving aside the controversial issue of whether having these properties has anything to do with having a soul.

So, let's consider the suggestion is that having a female gender identity depends on having certain key psychological characteristics. As I noted, the standard view holds that someone can have a female gender identity even if one lacks many typical feminine mannerisms, habits, and dispositions. Perhaps having a female gender identity supervenes on the psychological property of having a strong and persistent belief that you are a woman. But again, that merely pushes back the question: What exactly is it that you believe about yourself? This property you attribute to yourself – what does *it* supervene on? Perhaps there are some deeper aspects of personality that define one as a woman.

However, people that are uncontroversially women display a wide range of personality types and share various aspects of personality with many men. To develop the view that gender supervenes on psychology, we would need to address the following questions: Which psychological characteristics are masculine and which are feminine? How do we determine which psychological characteristics are key? Is it up to us how to define our concepts, or is there some independent fact of the matter? Are these characteristics fixed for all times and places, or do they vary from culture to culture? Being feminine has meant different things to different people at different times. We should also keep in mind that whatever features one specifies, be they emotional expressiveness, nurturing, or submissiveness, etc., these characteristics will admit of degrees, and are unlikely to be exclusive to women.

Another suggestion is that having a female gender identity depends on having a feminine brain. On the standard view, even biological males can have feminine brains, especially if their prenatal environment lacked sufficient testosterone. Transsexual.org *defines* a transsexual as 'a person in which the sex-related structures of the brain that define gender identity are exactly opposite the physical sex organs of the body' (Reitz, 2004). However, this suggestion is little more than an article of faith unless someone can tell us which structures of the brain define gender identity. Science has yet to demarcate clear brain differences between 'normal' men and women. The results of research on differences between male and female corpus callosi are sketchy (Fausto-Sterling, 2000). Other research focuses on a region of the hypothalamus known as the BSTc. Males are said to average twice as many neurons in this area as females. Some studies indicate that transsexuals have neuron numbers atypical of their biological sex, but typical of the sex that they believe should be (Zhou et al., 1997). However, the relationship between the BSTc and gender behavior is unclear. According to Christopher Horvath, 'There is as of yet no explicit neuropsychological theory that links brain structures to gender traits' (Horvath, 1999). Furthermore, one needs some characterization of gender traits and gender behavior before one can label the brain structures that cause those traits and behaviors 'gendered.' And furthermore, even if clear sex differences were found in the brain, we would still need to determine the causes of those differences: if brain differences were caused by differential environment conditions, we might see gender identity as a malleable product of gender socialization. And whatever the causes, any differences would be matters of degree, with a spectrum of variation ranging from highly feminine to highly masculine.

The claim that gender supervenes on biological, psychological, and/or social properties is very modest and plausible. However, the standard view has trouble accommodating this claim, for since proponents seem to hold that gender identity can vary independently of most biological, psychological and social properties. Furthermore, the idea that gender supervenes does not sit comfortably with a further tenet of the standard view, that every person has a gender identity that is either male or female. If one couples gender dualism with the view that gender supervenes, one is committed to the view that the vast array of human social, psychological and neurological properties sort themselves into two types. This is in tension with the fact that there is little consensus about which properties constitute the supervenience base for a female gender identity as apposed to a male gender identity, and also with the fact that the properties that are likely candidates for the supervenience base are possessed by individuals in a broad spectrum of varying degrees. So, part of the reason that the standard view is at odds with the claim that gender supervenes is because of the second claim about gender that I defend – that gender has no essence.

10.5. GENDER HAS NO ESSENCE

If gender supervenes, then having a masculine or feminine gender will depend on having certain physical, psychological, and/or social characteristics. But which characteristics? There is no simple answer. There is no single trait, or definite set of traits, had by all members of a gender, across cultures, throughout history. There is no single set of features that is necessary and sufficient for being masculine, or for being feminine. Gender is at best a matter of having enough of a number of different characteristics, to a sufficient degree, in a particular social context.

One way to put this point is to say 'gender has no essence.' I take it that for a kind to have an essence is for all and only members of that kind to necessarily share certain essential features. An essential feature of a kind is a property that a thing cannot lack whilst it is a member of that kind. Having four sides is an essential feature of squares. The kind 'squares' has an essence that consists in having four sides of equal length, etc. But some kinds have no essence. For those kinds, there is no set of features that all and only members share. Rather, they are grouped together for some other reason, perhaps family resemblance. Of course, we find this idea in Wittgenstein:

> Consider for example the proceedings that we call 'games'. I mean board-games, card-games, Olympic games, and so on. What is common to them all? – Don't say: 'There *must* be something common, or they would not be called "games"' – but *look and see* whether there is anything common to all. – For if you look at them you will not see something that is common to *all*, but similarities, relationship, and a whole series of them at that. ... And the result of this examination is: we see a complicated network of similarities overlapping and criss-crossing: sometimes overall similarities, sometimes similarities in detail.
> I can think of no better expression to characterize these similarities than 'family resemblances'; for the various resemblances between members of a family: build, features, colour of eyes, gait, temperament, etc. etc. overlap and criss-cross in the same way (Wittgenstein, 1958).

In the case of gender, there are no properties shared by all and only members of a particular gender (except for the gender property itself). However, we do find criss-crossing and overlapping similarities along various dimensions among the people we call 'women.'

When a concept is applied to individuals based on an indeterminate cluster of interrelated traits and family resemblances, it is a cluster concept. I suggest that gender concepts are cluster concepts. But note: cluster concepts do not typically have sharp boundaries. We categorize someone as masculine or feminine if they have enough of certain characteristics, to a sufficient degree. But how many is enough, and what degree is sufficient? There is no precise answer, for the concept is vague. If you could give a precise answer, you could state necessary and sufficient conditions for being a member of that kind. But since you can't, there is possibly an array of better to worse exemplars of that kind. Boundaries could be drawn in a number of ways, but there are no sharp natural boundaries to be found.

imagine having to sketch a sharply defined picture 'corresponding' to a blurred one. ... [I]f the colors in the original merge without any hint of outline won't it become a hopeless task to draw a sharp picture corresponding to the blurred one? Won't you then have to say: 'Here I might as well draw a circle or a heart as a rectangle, for all the colors merge. Anything – and nothing – is right.' And this is the position you are in if you look for definitions corresponding to our concepts in aesthetics or ethics (Wittgenstein, 1958).

And, I would add, this is the position you are in if you look for definitions corresponding to our concepts of gender. If gender concepts are cluster concepts, varying degrees of masculinity and femininity along different dimensions, and borderline cases, are likely.

It is questionable whether even biological sex has an essence. In order for a sex to have an essence, there would have to be some feature or set of features that every member of that sex has. But every feature that seems like a candidate for being definitive of sex faces counterexamples – individuals who lack that feature who we nevertheless want to categorize as members of that sex. While this deserves much more consideration that I will give it here, allow me to mention a few suggestive examples (Dreger, 1998; Fausto-Sterling, 2000).

Consider the suggestion that what is essential to being male or female is playing a certain functional role. The most feasible candidate is a biological role – a role in reproduction. But of course, many humans play no role in reproduction due to age, sterility, or lifestyle choice, and we do not decline to call them male or female. (Or, do we want to say that an infertile woman in still a woman in the sense that a broken clock is still a clock?)

Physical features, such as body size, strength, fat, and hair distribution generally differ between males and females. However, your stereotypical '98 pound weakling' is nevertheless male. There is much variation and a fair amount of overlap between male and female, and so these characteristics are unlikely to add up to a set of necessary and sufficient traits. Neither primary nor secondary anatomical sex characteristics can fully define sex. A castrated man is still male. Breast size varies significantly among women *and men*. Some females have a vaginal agenesis, and are born without a vaginal canal. In the 1800s, scientists and medical men thought that the essence of sex was the gonad (Dreger, 1998). If ever there was any doubt about the sex of an individual, ovarian tissue determined femaleness and testicular tissue determined maleness. However, after decades of being confronted with numerous cases of hermaphrodites who had the gonads of one sex, but the body type of the other sex, doctors revised their views.

One might then turn to hormones as the essence of sex difference. However, no type of hormone is exclusive to males or females. Each individual has a mix of hormones, and degrees of receptivity vary. Picking a particular mixture or threshold would seem to be an arbitrary decision when it comes to defining the essence of different kinds. Today, the tendency is to think that genes, particularly the 23rd pair of chromosomes, are the essence of sex. But consider someone with Androgen Insensitivity Syndrome – an XY individual whose body is not sensitive to testosterone. Such a person does not develop male genitalia or secondary male sex characteristics, but instead develops a body type typical of a female. Most people are reluctant to call such individuals male.

Perhaps certain features of the brain define sex. However, as far as we know, there is no anatomical structure in the brain of one sex that is totally lacking in other. One would have to compare differences in size of things like cross sections of the corpus callosum, or regions of the hypothalamus. However, differences in size and connectivity of various brain structures have multiple degrees of variation. Prenatal hormones influence brain development, but the precise mixture of prenatal hormones comes in more than two varieties. If brains can be feminine, it would stand to reason that some brains are more feminine than others. If being female means having a feminine brain, then it is a vague region of a continuum, rather than a discrete category exclusive to women.

I'm not saying that 'male' and 'female' are fictions, or even social constructions. By and large, people tend to have a certain characteristics that easily allow us to categorize them as either male or female. However, there is no particular set of characteristics that each person must have in order to be male, or female. The concepts of male and female are cluster concepts, like the concept of a game.

If we can't define the essence of biological sex, how much less of a chance do we have of defining the essence of gender? There is no certain set of personality traits, such as being nurturing, intuitive, or sensitive, that is necessary and sufficient for being feminine. Even more so than hormones and features of the brain, psychological traits are had in various combinations and degrees, with significant overlap between men and women. How else could men get in touch with their 'feminine side'? Every person has a unique mix of psychological characteristics. It would follow that some people have a female gender identity to a certain extent, while others have a female gender identity to lesser extent. The division of gender into two types would seem to be arbitrary.

One might suggest that we can define the essence of feminine gender by forming a disjunction of all of the sets of traits that are sufficient for having a female gender identity. However, it seems that gender concepts are too malleable and open-ended to permit such an analysis. The concept of femininity has not remained constant over the centuries. It is quite plausible that some future person will exemplify some unanticipated cluster of traits that will qualify her as feminine. Or, to quote Wittgenstein again:

But if someone wished to say: 'There is something common to all these constructions – namely the disjunction of all their common properties' – I should reply: Now you are only playing with words. One might as well say: 'Something runs through the whole thread – namely the continuous overlapping of those fibers'(Wittgenstein, 1958).

Because our gender concepts are cluster concepts with no precise boundaries, they are difficult to operationalize in any scientific way. In his paper 'Measuring Gender,' Christopher Horvath examines a number of studies that have tried to establish a correlation between childhood gender non-conformity and adult homosexuality. In order to do so, researchers must determine when and to what extent an individual exhibits gender behavior. But there is no consensus about how to do this. Horvath notes:

the aspects of behavior and physiology that researchers treat as markers of 'gender' (rather than sex, class, or culture) differ greatly from study to study and from scientist to scientist.
they [the scientists] provide no uniform, consistent method for identifying and measuring the biologically significant components of gender.
particular combinations and degrees of the attributes, interests, attitudes, behaviors, etc. manifested may substantially vary between people with equally strong, unambiguous gender identities...

Horvath concludes 'gender-typical phenomena are multi-dimensional and multi-factorial' (Horvath, 1999). Furthermore, certain components of gender, such as preference for certain kinds of attire or occupations, are unlikely to be directly based in biology. Some aspects of gender and gender role are culturally relative. One cannot define a kind of social role that all and only women have, historically, cross-culturally, or even within modern societies.

So, what is the person with cross-gender identification identifying with? A multi-dimensional, multi-factorial cluster of culturally relative psychological traits? Perhaps. Let me put the question another way. The biological male with GID claims to have the feelings and reactions typical of women. But what feelings and reactions *are* typical of women? Many people probably have more in common, in terms of feelings and reactions, with members of their own social group, such as their family, church, peer group, occupation, class, or culture, than they do with members of their own sex in different social groups. Perhaps the person with GID only identifies with the opposite sex within his or her society. Different societies have somewhat different gender norms, so one personality type might identify as female in one society, but as male in another. But, if it is a consequence of social factors that only women typically have a particular set of feelings and reactions, then to insist that only someone with a female body type should have those feelings and reactions is to valorize the gender norms of that society. Even if biological factors determine that women typically have certain psychological characteristics, to insist that someone with these psychological characteristics should have a female body type is to stand as a defender of the norm.

The American Psychiatric Association claims that it plays no such a role. According to the DSM-IV, 'Neither deviant behavior ... nor conflicts that are primarily between the individual and society are mental disorders unless the deviance or conflict is a symptom of dysfunction.' But interestingly, the manual says 'GID can be distinguished from simple nonconformity to stereo-typical sex role behavior' not because it is a symptom of a dysfunction but 'by the extent and pervasiveness of the cross-gender wishes, interests, and activities' (Task Force on DSM-IV and American Psychiatric Association, 2000). Apparently, if the deviance is severe enough, it does count as a disorder. And what could be more severe than repudiation of a society's gender norms?

10.6. CONCLUSION

Admittedly, the world contains masculine men, and feminine women, but these categories are not exhaustive, exception-less, immutable, or clearly defined. Different societies have had different standards for categorizing people according to gender. Whether an individual is considered a man, woman, or something else, in a society depends upon the standards of that society, and whether the individual has the physical and psychological features to satisfy that those standards. In societies with exactly two well-defined gender norms, individuals feel pressured to exemplify one cluster of characteristics, to the exclusion of the other. A list of characteristics is difficult, if not impossible to articulate in any great detail, but it typically includes primary and secondary sex characteristics, modes of dress and grooming, personality, preferences, occupations, expectations, and relationships. While we can argue about how 'natural' it is for most people to exemplify one cluster of characteristics to the exclusion of the other, clearly a significant number of individuals (conservatively 10,000 in the U.S.) find it difficult to exemplify an acceptable cluster.

While transsexuals may seem to challenge gender norms, in a sense, they embrace them. The desire to change one's body to match one's perceived gender identity reveals acceptance of the idea that sex and gender must coincide, that certain behaviors and desires are incompatible with certain physical characteristics. The transsexual does not reject the gender roles of his society; he merely rejects one gender role in favor of another (Fausto-Sterling, 2000). MtFs do not typically object to stereotypes of women – they want to personify those stereotypes, and often display an exaggerated femininity. But perhaps this says less about the transsexual's 'gender identity' and more about her society.

Most societies assign individuals to one of two possible genders, and which gender you are assigned determines the character of your interactions with others and your life prospects in countless ways. In some societies, if one is uncomfortable with one's gender role, there is exactly one other option – the opposite gender role. But to succeed in that role, one must look the part. For some people, physical alteration is their best chance of conforming to an available and acceptable gender option. While the standard view sees the pre-operative transsexual as a person with a conflict between mind and body, the FtM philosopher Adam Parascandola wonders '... is it that my internal self is in conflict with *society's view* of my external body? ... I often wonder, if society did not insist on granting identity based on external characteristics, whether I would have felt the need to change my body' [(Parascandola, 2001) my emphasis]. If it weren't for the fact that so many facets of one's life are largely determined by the gender that one appears to be, would there be a need to change bodies to 'match' minds? I suspect not. I suspect that the phenomenon of cross-gender identification has more to do with a broad range of personality types trying to cope with a rigid two-gender system than it does with 'crossed wires' or souls that end up in the wrong kind of body. Hence, I suggest that so called 'GID' is primarily a conflict between the individual and her society, and only derivatively a conflict between the individual and her body. Greater social tolerance of gender diversity could create a context within which such individuals would not be considered 'disordered.'

University of Nebraska-Lincoln, Lincoln, Nebraska, USA

REFERENCES

2002. Changing sexes: female to male. Documentary
2004a. Harry Benjamin International Gender Dysphoria Association. Available from http://www.hbigda.org/socv6.cfm
2004b. The opposite sex: Rene's story. Documentary
Conway Lynn (2002) How frequently does transsexualism occur?' Available from http://ai.eecs.umich.edu/people/conway/TS/TSprevalence.html
Dreger Alice Domurat (1998) Hermaphrodites and the medical invention of sex. Harvard University Press, Cambridge, MA
Fausto-Sterling Anne (2000) Sexing the body: gender politics and the construction of sexuality. Basic Books, New York
Horvath Christopher (1999) Measuring gender. Biology and philosophy 14:505–519
Kohlberg LA (1966) Cognitive-developmental analysis of children's sex-role concepts and attitudes. In: MacCoby EE (ed) The development of sex differences, Stanford University Press, Palo Alto, CA
Meyerowitz Joanne (2002) How sex changed: a history of transsexuality in the United States. Harvard University Press, Cambridge, MA
Money John (1976) Prenatal and postnatal factors in gender identity. In: Serban George, Arthur Kling (eds) Animal models in human psychobiology, Plenum Press, New York
Parascandola Adam Linda (2001) Trans or Me? In: Kolak, Daniel, Raymond Martin (eds) The experience of philosophy, Oxford University Press, New York
Rees Tamara (1959) Male becomes female. Sexology 26:212–218
Reitz Jennifer Diane (2004) Transsexuality. Available from http://transsexual.org/
Task Force on DSM-IV, American Psychiatric Association (2000) Diagnostic and statistical manual of mental disorders DSM-IV-TR (Text revision), 4th edn. American Psychiatric Press, Washington, DC
Wittgenstein Ludwig (1958) Philosophical investigations. Macmillan, New York
Zhou JN, Hofman MA Gooren LJ, and Swaab DF (1997) A sex difference in the brain and its relation to transsexuality. The international journal of transgenderism 1(1)

11. CLINICAL TRIALS AS NOMOLOGICAL MACHINES: IMPLICATIONS FOR EVIDENCE-BASED MEDICINE

11.1. INTRODUCTION

Evidence-based medicine (EBM) has been defined as 'the conscientious, explicit, and judicious use of current best evidence in making decisions about the care of individual patients' (Sackett et al., 1996). In practice, 'current best evidence,' at least for treatment decisions, is generally taken to be the results of randomized controlled trials (RCTs). In this paper, I examine this notion of best evidence. I will begin by arguing that clinical trials are an example of what Nancy Cartwright has called 'nomological machines.' That is, they require the development of a set of circumstances designed to give replicable results if the experiment is repeated over time. They are not, however, designed to inform clinical practice. Because of this, the incorporation of the results of a clinical trial into practice requires some rather complex analysis. I will then illustrate this process of analysis by extending Cartwright's metaphor of a 'nomological machine' and showing that to understand how the results of a clinical trial can inform clinical practice requires that the machine be 'reverse engineered.' However, even the best techniques of evidence-based medicine can usually provide only rudimentary tools for such reverse engineering.

Having discussed the limitations of nomological machines, I will return to Cartwright's work in order to launch a deeper critique of RCTs. Cartwright's own discussion of nomological machines draws on their use in economics and, more importantly, in physics, where a properly constructed nomological machine can give us knowledge of fundamental physical processes. This type of knowledge, however, is far from the kind we need from medical research. In medicine, what we want to know is the likely effects of a treatment in the case of an individual patient. This requires knowledge, not of fundamental processes, but of the ways in which the effects of the treatment are different in different circumstances. Drawing on Cartwright's own discussion of 'capacities' or 'natures,' I will show that RCTs, as nomological machines, provide only a very specific type of information about the capacities of the drug being studied, and that the increasing popularity of meta-analysis as a source of 'evidence' for clinicians indicates that these limitations are not recognized by the proponents of EBM.

11.2. THE DESIGN OF RANDOMIZED CONTROLLED TRIALS

The RCT is often described as the 'gold standard' of medical research. In the 'hierarchy of evidence' proposed by evidence-based medicine, it is said to provide the strongest evidence of the utility of a particular drug. Its purpose is to compare

149

H. Kincaid and J. McKitrick (eds.), Establishing Medical Reality, 149–166.
© 2007 *Springer.*

the effects of a particular treatment with those of a placebo or of another active treatment, and to do so under circumstances that have been carefully controlled in order to minimize bias, or confounding, by other factors that may influence the outcome of the trial. (In the following account, I will describe the simplest form of clinical trial: a two-arm, placebo-controlled trial with equal numbers of subjects in each group. The essential points I want to make are not importantly different for more complex trial designs.)

In order to minimize confounding, investigators set stringent criteria that must be met by subjects before they can be enrolled in a trial. Generally, these include controlling the age of patients, ensuring that patients' disease status is comparable, and eliminating potential participants with comorbid disorders or those taking other medications. As participants are recruited, they are randomized to receive either active medication or placebo. This step has the effect of controlling for unknown differences among participants; it is assumed that the process of randomization will balance these confounding factors between treatment groups. (Since this assumption, which is itself controversial, is made with regard to large numbers of repetitions of the same experiment, rather than to any single experiment, it is common practice for researchers to look for imbalances between groups after randomization.) Random allocation of subjects to treatment groups is also an assumption underlying the statistical analyses used in the majority of clinical trials.[1] In addition, wherever possible, 'blinding' is used to eliminate potential bias in assessing study outcomes; neither the subjects themselves nor the study personnel should know whether an individual is receiving active drug or placebo. While such knowledge presents less of a problem with 'objective' outcomes – such as quantified results on laboratory tests, in the case of outcomes for which subjective assessments of improvement must be made by participants or by those conducting the study, knowledge of the group to which an individual has been allocated may influence these assessments.

The RCT also requires subjects and experimenters to follow precise protocols; these are designed to ensure that all participants receive a uniform treatment. The drug must be taken in specified amounts, at specified times and, in some cases, under specified circumstances (with or without food, for example). The effects of the drug are also generally measured using standardized, usually quantifiable parameters, since individual differences in interpreting rating scales may be another source of bias. In sum, as Jadad notes, these trials are generally 'designed in such a way that the results are likely to yield a 'clean' evaluation of the interventions (1998, p. 12).

But what does a 'clean' evaluation mean? In one sense, it means that a successful intervention performs better than placebo at a predetermined level of statistical significance, and thus gives an estimate of the probability that observed differences in average outcomes between the two study groups have occurred in a population in which the measured effects of the drug are no different than those of the placebo. In the sense important to those interesting in practicing evidence-based medicine, however, it means that the results of a clinical trial cannot easily be extrapolated to the messy world of clinical medicine. From the first perspective, which could be called the 'statistical perspective,' a clean evaluation means that the trial has internal validity. Any differences in performance between the two interventions can (barring the existence of confounding factors that have a high correlation with the experimental intervention) be attributed to real differences in the effects of the two substances being compared, rather than to chance. (This conclusion assumes that the trial has been well-designed and provides adequate control, whether through the study design or through randomization, of known and unknown factors that may influence outcomes in the two groups and is not an inference that can be drawn directly from a statistically significant result.) This perspective, however, rests on what Goodman (1999) has called the 'p value fallacy,' which is the belief that the statistical significance of the results of an experiment give both the long-term outcomes of an experiment and the evidential meaning of a single experiment. According to Goodman, this fallacy stems directly from the common practice of conflating the statistical methods of Fisher with Neyman-Pearson hypothesis testing. The fact that the p value is believed to play this dual role, Goodman further argues, is the reason it is so common to interpret the results of a clinical trial without considering the other evidence bearing on the effectiveness of the drug. It is assumed, in other words, that any drug that has been shown to be effective in a clinical trial will be effective in clinical practice.

Although there are real grounds for worry that some individuals do commit this fallacy, the problem I want to address is more complex than that. Clinicians interested in developing skills in evidence-based practice have developed highly sophisticated techniques of critical analysis and are not likely to be easily fooled by p values. The *Users' Guide to the Medical Literature*, a standard evidence-based medicine reference text, counsels clinicians wishing to apply the results of a clinical trial to the care of their own patients. It notes that many patients seen in clinical practice would not have met the strict inclusion and exclusion criteria for the particular clinical trial, thus making it unclear whether it is reasonable to expect that any individual patient receiving that medication will respond similarly to the participants in the RCT. Moreover, even if the patient does share relevant characteristics with the study participants, the results of most RCTs are reported as average results; in reality individual subjects' responses to a treatment will vary, often widely. Thus, '[applying] average effects means that the clinician will likely be exposing some patients to the cost and toxicity of the treatment without benefit.' (Guyatt et al., 2001). The suggestions made for applying the results of a given study in clinical practice have to do with the assessment of the external validity of a study. However, even the *Users' Guide* does not seem to be aware of the extent of the difficulties inherent in this assessment.

All of this is not to suggest that RCTs are useless. Yet despite my earlier optimism about the rarity of the p value fallacy among clinicians serious about EBM, there are indications that even skilled practitioners of EBM mistake the purpose of RCTs. Quite simply, RCTs are *not* designed to inform clinical practice. Rather, they are designed to ensure that, on balance, the benefits of taking the drug outweigh the harms, and that the benefits are not due solely to a placebo effect. RCTs provide a standardized situation within which the effects of a drug can be studied and, as such, are examples of what Nancy Cartwright describes as 'nomological machines.' In the next section, I will discuss Cartwright's conception of a nomological machine and its implications for the interpretation of the 'clean' results of randomized trials.

11.3. RCTS AS NOMOLOGICAL MACHINES

On Cartwright's view, the goal of science is not, as traditionally argued, to discover the 'laws of nature.' (On this traditional account, it might be claimed that the goal of a clinical trial is to discover the impact of a particular therapy on individuals with a particular disease, and to express this impact in statistical terms.) Laws of nature are understood as regularities; in the terms of the Humean empiricism that has characterized much of contemporary philosophy of science, Cartwright says that a law of nature is 'a necessary regular association between properties' (1999, p. 49).[2] In some cases, the world itself is constructed in such a way that the laws governing it are naturally apparent (the example Cartwright gives here is of the planetary system). More often, however, the circumstances that enable scientists to uncover laws are constructed in the laboratory. Cartwright calls the creation of these circumstances the construction of a nomological machine, that is, of a machine that produces 'laws of nature.' She argues, 'the principles we use to construct nomological machines or to explain their operation can not adequately be rendered as laws in the necessary regular association sense of 'law' ' (p. 50).

Despite claiming that the majority of nomological machines occur only in experimental situations, Cartwright uses her example of a naturally occurring nomological machine, the motion of the planets, to illustrate the concept. She cites 'Kepler's problem,' which requires that the regular motion of the planets be accounted for 'using descriptions referring to material bodies, their states of motion and the forces that could change them ...in my terminology, the task is to figure out the nomological machine that is responsible for Kepler's laws. This means that we have to establish the arrangements and capacities of mechanical elements and the right shielding conditions that keep the machine running properly so that it gives rise to the Kepler regularities.' (1999, p. 50).

Cartwright goes on to show that Newton solved this problem by determining the magnitude of the force that would be required to keep a planet in an elliptical orbit: $F = GmM/r^2$, where G is the gravitational constant, m and M refer to the masses of the planet in question and the sun, respectively and r to the distance between them. The naturally occurring nomological machine that is the planetary system exhibits a regular (law-like) behavior described by Kepler and explained (by appeal to specific bodies and their motions) by Newton. The law, however, describes only the cases in which the ellipse is perfect. In cases where the observed motion of a planet deviates from an ellipse, it is because of the force exerted by a third body (a planet near enough and of a great enough mass) that it, too, exerts a force on the planet whose motion is to be explained. This brings us to another characteristic that Cartwright claims is essential to the operation of a nomological machine, shielding.

She points out that, prior to the discovery of Neptune, its existence was suspected because the observed orbit of Uranus deviated from that which was predicted by Newtonian principles. Thus, 'it is not enough to insist that the machine have the right parts in the right arrangement; in addition, there had better be nothing else happening that inhibits the machine from operating as prescribed' (ibid., p. 57). Because the two-body system described by Newton's law was not shielded from the effects of a third body, the original explanation of the behaviour of the nomological machine was inadequate.

Yet even though its motion could not be predicted on the basis of Newton's original model, the orbit of Uranus *is* regular. Accounting for it requires a revision of the original nomological machine proposed by Newton. In the vast majority of cases, however, the phenomena we wish to describe are not so regular. In these cases, we must *construct* a nomological machine that makes evident the right sorts of regularities. In the case of clinical trials, the regularities in which we are interested are the effects of the drug being tested (its impact on outcomes reflecting disease or symptom severity as well as additional 'adverse effects' that are caused by the drug). Creation of such a nomological machine requires that we bring the right parts together in the right arrangement and that we shield these pieces of the machine from other effects.

Cartwright defines a nomological machine as 'a fixed (enough) arrangement of components, or factors, with stable (enough) capacities that, in the right sort of stable (enough) environment will, with repeated operation, give rise to the kind of regular behavior that we represent in our scientific laws.' (ibid). The relevant components in a clinical trial are the interventions – the drug of interest and its comparator (active or placebo), the patients, and the rules governing their relationship (the dosage and the length of time for which the drug is taken, the timing and method of measurement of outcomes). The interventions themselves are, of course, standardized, and the patients are standardized as well. This 'standardization' of patients reflects the need for shielding to ensure that nothing is happening 'that inhibits the machine from operating as prescribed' (Cartwright, p. 57). In clinical trials, shielding occurs by controlling the sorts of potential confounding variables described earlier, thus ensuring (so far as possible) a clean result from the trial.

In Cartwright's terms, then, the arrangements of the components of the nomological machines are stable (enough); the capacities of both the interventions and the patients are held to be stable and any other factors that might affect the physiological response of the patients (their capacity to respond) to the intervention are controlled for as much as possible, either through inclusion and exclusion criteria or, where necessary, through documentation during the course of the trial and, in extreme cases, through the withdrawal of patients from the trial or their exclusion from the analysis of results.

There is also a second way in which clinical trials are an example of Cartwright's nomological machines. Cartwright argues that the only reason that we can use nomological machines to produce laws of nature is that we *construct* the machines in such a way as to produce them. The components of the machine are chosen and assembled to give an answer to a specific question, and that answer is a 'law of nature.' 'We get no regularities without a nomological machine to generate them, and our confidence that *this* experimental set-up constitutes a nomological machine rests on our recognition that it is just the right kind of design to elicit the nature of the interaction in a systematic way.' (p. 89). Cartwright suggests that what a proper nomological machine actually does is to ensure that the capacities possessed by its components are revealed in a certain way. I will return to this point later; for now, however, I want simply to emphasize that, on Cartwright's account, it is these capacities that are basic, rather than the 'law-like' regularities in their effects. Newton's law of gravitation, for example, tells us about the force between two bodies, where force must be understood as an abstract term 'that describes the capacity of one body to move another towards it, a capacity that can be used in different settings to produce a variety of different kinds of motions' (p. 52). It is a mistake to consider the regularity or regularities to which a capacity gives rise when it is exercised in the context of a particular nomological machine to be the 'real' effect of that capacity, and the regularity to which it gives rise as the 'real' law describing its effects.[3] Had we combined the capacity of interest with different capacities, or changed the shielding conditions, different behaviour and a different law would have resulted. Recall Cartwright's example of planetary motion; a different law than that of a three-body system describes the motion of a two-body system. Even though the capacity to attract other bodies (possessed by all bodies in the system) is the same in both cases, it is manifested differently in different contexts. In the case of clinical trials, the mistake often made in interpreting their results is to take the 'clean' results of the operation of the nomological machine for the 'real' effect of the drug, rather than the manifestation of the drug's capacities in a particular set of circumstances. This set of circumstances, remember, was designed to standardize the context in which the drug and the placebo are compared, not to mimic the complexities of clinical practice.

Clinicians have made similar observations about the goal of RCTs. Moyé (2000) suggests that the purpose of RCTs is to prevent the introduction of 'noxious placebos' into clinical use. All drugs, he points out, have side effects, and if the efficacy of a medication is not significantly higher than that of a placebo, the drug will do more harm than good. Thus, the design of the RCT is meant to facilitate comparison of the study drug and the placebo. Wulff et al. (1996) draw a distinction between biological medicine (basic scientific research) and clinical research (controlled trials). They suggest that clinical research is more akin to technology than to science – putting a drug through a clinical trial is like testing an airplane in a wind tunnel. They do, however, accept that some conclusions about the drug's effect in clinical practice may be drawn; on the basis of a clinical trial with a 95% confidence interval of 15–45%, a clinician is justified in accepting the 'statistical law' that between 15 and 45% of patients will benefit from this drug. While this conclusion is in reality another form of Goodman's *p* value fallacy, since *p* values and confidence intervals are mathematically related, the use of confidence intervals does have the pragmatic advantage of drawing attention to the fact that further interpretation of the results of a clinical trial are required. Even so, Wulff et al. gloss over the question of how closely patients must resemble those subjects in the trial for this generalization to be feasible. As discussed earlier, additional information is required for a clinician to determine whether her individual patient is in that 15–45% of the population. Here again, Cartwright's description of nomological machines is helpful.

11.4. TRIAL INTERPRETATION AS REVERSE ENGINEERING

Once we realize that a clinical trial is a nomological machine, designed to allow us to answer basic questions about the effects of a therapy relative to a comparator, then the next step is to recognize that the results of such a trial cannot be interpreted without an understanding of how the machine works. In the case of clinical trials, this involves considering the results of the trial in the broader context of biology. In the terms of Wulff et al. (1996), it requires moving back from clinical research into biological medicine. Only then can the nomological machine become helpful to clinicians. To extend Cartwright's metaphor, then, I propose that interpreting the results of a clinical trial requires a strategy known as 'reverse engineering.' This strategy is described by Dennett (à propos of a different set of philosophical problems):

When Raytheon wants to make an electronic widget to compete with General Electric's widget, they buy several of GE's widgets and proceed to analyze them: that's reverse engineering. They run them, benchmark them, X-ray them, take them apart, and subject every part of them to interpretive analysis. Why did GE make these wires so heavy? What are these extra ROM registers for? Is this a double layer of insulation, and, if so, why did they bother with it? Notice the reigning assumption that all 'why' questions have answers. Everything has a *raison d'etre*; GE did nothing in vain (1995, p. 212).

In the case of reverse engineering RCTs, there is an additional challenge; as well as the usual task in reverse engineering of looking at the specific choices made by its designers and analyzing the reason behind them, the engineer also has to notice what is missing from the trial – what choices have not been made – and to consider the implications that different choices in the design of the trial might have had on its outcome.

To a certain extent, the skills required for reverse engineering are precisely what is taught by EBM. The *Users' Guide*, for example, contains explicit discussions of how to use the results of various types of clinical research (including RCTs) in clinical practice. The *Guide* explains how to find and critique relevant studies by applying various methodological criteria, including assessments of the quality of the randomization and blinding techniques reported in the published trial results. Secondly, a clinician must consider whether the trial has shown the drug to be effective at all, by, for instance, calculating the '*Number Needed to Treat*'[4] in order to prevent one adverse outcome (death, heart attack, etc.) or to effect one good one. The lower the number needed to treat, the more benefit that can be expected from the intervention.

Such lessons in the practice of EBM equip physicians to wring the maximum amount of information out of the clinical trial. It should further be noted that one of the successes of EBM has been to improve the quality of reporting in clinical trials so that physicians are more easily able to extract the information they require. However, even the best (in terms of methodological and reporting standards) clinical trials are not informative enough. While EBM does teach physicians how to critically appraise clinical research, it is less help in applying these lessons in clinical practice. The *Users' Guide* does give advice as to how to do so, but it is of limited utility. Having completed the first two steps described above, a clinician is supposedly ready to consider the benefits of the drug for her patients in particular. Here, she is supposed to ask herself whether the patients included in the study are similar to her patient, whether all clinically important outcomes were considered in the trial and whether the benefits to be expected from this therapy would be worth the expected harms and costs to the patient (Guyatt et al., 2001). While these are important and appropriate questions, the *Users' Guide* does not have a great deal of advice as to what to do when the answer to them is 'no.' Rather, they suggest that, in answering the first question, 'a better approach than rigidly applying the study's inclusion and exclusion criteria is to ask whether there is some compelling reason why the results should not be applied to the patient. A compelling reason usually will not be found, and most often you can generalize the results to the patient with confidence' (ibid., p. 71). While this may be true, it certainly seems contrary to the spirit of EBM; moreover, it ignores the fact that the particular inclusion and exclusion criteria used in a trial are carefully chosen in order to eliminate those factors that the designers of the trial have reason to believe *will* affect the drug's efficacy in the study patients. There seems no reason, then to assume that they won't have much impact on the drug's effect on individual patients seen by clinicians. Finally, even patients who would have met the inclusion and exclusion criteria for a study may not benefit from a particular drug. As mentioned earlier, clinical trials report the average results obtained from the experimental and the control groups for each outcome measured. However, they do not tend to speculate as to what factors distinguish the trial patients who failed to benefit from those who did (or, for that matter, to examine factors that increase patients' benefits from a treatment). Yet there may be good reasons that can be found to explain these outcomes.

In a study examining the reasons for treatment failure in a series of patients with onychomycosis (a fungal infection of the nails), Gupta and Kohli (2003) identified a number of factors peculiar to these patients that might account for treatment failure. None of these factors is one of the standard exclusion criteria in clinical trials of medications for onychomycosis. Generally, these studies exclude, for example, patients who are diabetic (they tend to have poor peripheral circulation, which lessens the amount of drug that reaches the affected site), individuals with compromised immune systems (whose ability to 'fight off' the pathogen is thus impaired), as well as people taking concomitant medications that may interfere with the action or the metabolism of the study drug. There is no reason to believe, then, that the patients examined in the Gupta study would not have been enrolled in a clinical trial of either terbinafine or itraconazole. While there is some difference in their specific mechanisms of action, both of these drugs are antifungal agents that kill the fungus that causes the infection. Both have been shown, in a number of clinical trials, to be effective treatments (see, e.g. Crawford et al., 2002). Yet all of the patients in the Gupta study failed to benefit from these standard treatments; fourteen had been treated with both terbinafine and itraconazole and four with terbinafine only. However, in seventeen of the eighteen patients studied, one or more factors were present that may have explained treatment failure: in some, the clinical presentation of the infection was of a nature that had been previously associated with difficult-to-treat disease (the presence of a lateral streak or a longitudinal streak of infection in the nail or of a dermatophytoma, which is a ball of fungus not easily penetrated by the medication), or simply a very severe case of infection, as measured by the percentage of the nail involved. Several patients had very thick nails, or previous trauma to the nail, both of which also reduce the amount of drug that reaches the site of infection. Finally, one patient was obese, with poor peripheral circulation.

It is important to note that all of these potential explanations for treatment failure were determined only after the majority of the patients had undergone multiple courses of treatment. Both terbinafine and itraconazole are oral medications, thus their effects are exerted on the entire body, rather than just on the site of infection. Both also require treatment for an extended length of time, several months for most treatment regimens. Thus, these patients were exposed to the toxicity (and the expense) of treatments that failed; however, since these treatments are the standard of care for this disorder and have been shown to be effective in clinical trials, it is rare that a physician treating a patient for this disorder will not turn to these medications first. While this was not its intent, the Gupta/Kohli study provides examples to demonstrate that clinical trials are not sufficient guides to treatment. Nor can 'reverse engineering' a trial show a physician that treatment was likely to fail in these cases. Since clinical trials rarely discuss the reasons for treatment failure in the patients who do not benefit from the study drug, it is also possible that some of the patients in the study shared the characteristics of the patients in the Gupta study; this information was not, however, available to physicians seeking to interpret the results of the trial.

I have reviewed several factors that may limit the ability of physicians to reverse engineer a clinical trial, including the standards governing the design of clinical trials, and the practices followed in their reporting. Moreover, these are limitations that even the best practitioners of EBM cannot overcome. In the following section, I will consider in greater detail the issues in the design of clinical trials that limit their utility in clinical practice.

11.5. BEHIND THE SCENES OF THE RCT

I have previously noted the characterization given by Wulff et al. (1990) of clinical trials as analogous to testing an airplane in a wind tunnel. This characterization is compatible with my suggestion that clinical trials are a species of Cartwright's nomological machines. Airplanes are built on the basis of knowledge drawn from such sciences as aerodynamics and classical mechanics. So, too, are drugs developed based on knowledge derived from numerous disciplines, including physiology, pharmacology, biochemistry and pathology. When the time comes to test the airplane in the wind tunnel, or the drug in an RCT, all of this knowledge is bracketed, or treated like a 'black box,' for the purposes of the test. The differences between the two situations lie in the response to a successful outcome. In the case of the airplane, a successful outcome means that the plane is safe and suitable for use. The 'black box' need not be opened again. This is because it is taken to be a reasonable assumption that the wind tunnel sufficiently mimics the 'real' environment in which the plane will be expected to perform. Thus, passing the test in the wind tunnel means that the plane will perform appropriately in its real environment.

This is not the case, as I have argued above, with the testing of a drug in a clinical trial. Unlike the wind tunnel, the clinical *trial* environment is not an adequate model of the *clinical* environment in which the drug is expected to perform. Thus, the need for a type of reverse engineering that looks at the options or parameters that have *not* been selected in designing the trial and at how selecting those options might have altered the outcome of the trial.

This aspect of reverse engineering can be better understood by returning briefly to Cartwright's discussion of nomological machines. She notes that her interest in the concept of a nomological machine was prompted by her work on models in physics and in economics. She describes models as 'blueprints for nomological machines.' That is, models provide the conceptual tools that allow researchers to build nomological machines. The important point here for my argument is that in most instances, and certainly in the case of clinical trials, the models are developed on the basis of 'concepts from a variety of disciplines' (Cartwright, 1999, p. 58). Thus to build, or to reverse engineer, a nomological machine is inherently multidisciplinary. What is required then, in the case of RCTs is an understanding of what each of the 'pieces' of the nomological machine is doing (why certain types of subject were enrolled, why certain endpoints were chosen, what the statistical analysis is testing), and how the result of the running of the machine depended on this combination of pieces. This process requires simultaneous consideration of information from all of the disciplines that contribute to trial design; none on their own is enough to provide an adequate interpretation. However, this is precisely the information that is 'black boxed' in an RCT – and in the published report of the study. In other words, neither the statistical analysis nor the discussions of the results of a trial make reference to this background information. This information is not available to the would-be practitioner of EBM.

Protocol

The decision to conduct an RCT for a particular medication in an RCT is based on a number of factors that lead scientists to believe that a full-scale RCT is warranted. By the time an RCT of the type I have described is conducted, the drug has been tested in a number of preclinical and clinical experiments. Preclinical research may involve animal studies that provide information about the physiological mechanisms by which the drug is metabolized and clues as to what its effects may be in human beings. The first clinical studies conducted in human beings (known as Phase I trials) are conducted only when there is evidence of the safety, and the potential efficacy, of the drug from animal research. Phase I trials are designed to measure the safety and metabolic effects of the drug at different doses and so, on the hierarchy of evidence, correspond to a case series in their methodology (Jadad, p. 15), since there is no control group. In trials of drugs for many conditions, though, Phase I studies are conducted on healthy volunteers and so provide no evidence for the efficacy of the drug in the treatment of a disease.

Once a drug has been shown to be effective in Phase I trials, testing can progress to phase II trials. In these trials, there is generally no control group, but the drug is given to patients, rather than to healthy volunteers. These trials tend to be small, with about twenty patients participating (Jadad, p. 15) and their aim is to test the efficacy of varying dosages and frequencies of administration. On the basis of the results of these trials, the treatment regimens for Phase III trials – the full-scale RCTs described earlier – are chosen.

Despite the close relationship between these earlier studies and the RCT whose design they inform, the results of Phase I and Phase II trials are rarely published in medical journals, and so are not available to clinicians who want to engage in reverse engineering. Moreover, the Phase III trials that are published, and are the bedrock of EBM, do not supply this missing information, whether to link the current study with preclinical work or to earlier trials (Phase I to III) of the same drug. In most studies, the description of the trial is generally limited to the design and analysis of the trial itself without any discussion of context. The discussion sections of the average published RCT have been described as 'islands in search of continents' (Clarke and Chalmers, 1998) because they fail to take into account 'the totality of the available evidence.' Of the 26 trials analysed by Clarke and Chalmers, 19 demonstrated 'no evidence' that an attempt had been made to integrate the results of the trial with those of previous trials. Similarly, a positive ('significant') result is not likely to be critically assessed in light of other evidence bearing on the efficacy of the drug. (In part, this is because publication practices make it expedient to emphasize the significance of results – negative trials are rarely published. Detracting from the 'significant' p value in order to provide a more balanced assessment of the trial is like looking a gift horse in the mouth).

In summary, the use of clinical trials as 'evidence' to inform clinical practice faces problems on a number of levels. First, it is only through reverse engineering a clinical trial that its utility in a particular clinical situation can begin to be assessed. However, the very purpose of a clinical trial is to strip the experiment itself of any context; it is simply a nomological machine. In addition, much of the information on which the trial is based is unavailable to those reading the published report of the trial. This means that, in practice, only a limited kind of reverse engineering can be undertaken.

But there is a deeper reason that clinical trials, as nomological machines, are only incomplete resources for EBM. This reason has to do with the nature and goals of research in biology and medicine. In the next section, I will return to Cartwright's discussion of the relationship between laws and capacities in order to elucidate the nature of research in biology and medicine.

11.6. LAWS AND CAPACITIES IN PHYSICS AND IN BIOMEDICAL RESEARCH

In the preceding sections I have suggested that clinical trials are an example of what Nancy Cartwright describes as 'nomological machines.' However, Cartwright's scientific interests (though perhaps not her philosophical ones) are quite distinct from mine; she is concerned primarily with work in what she calls exact sciences, such as physics and economics, which are highly mathematical and also tend to require rigorous derivation of conclusions from theories. Her concern is to show that these derivations are possible because of the structure of the theories, because both physics and economics use models that allow the building of nomological machines. It is through the use of appropriate nomological machines that scientists are able to isolate specific capacities of an object being studied and to develop formulations of the laws governing the expression of that capacity (at least in the context of the operation of the nomological machine). Cartwright's early work on laws, in *How the Laws of Physics Lie* (1983), argues against 'fundamentalists' who take these laws to be capturing the real behaviour of the object, or the essence of the capacity, being studied. What laws actually do, she claims, is to express their behaviour in a certain, highly controlled, context.

In applying Cartwright's analysis to the case of clinical trials, we must first recognize that the types of capacities described in physical laws are different from those studied in biology. This is the reason that biomedical science is not an 'exact science' (in Cartwright's terms); it is neither mathematical nor derivational. Rather, it operates primarily at the level of elucidating causes, or qualitative or quantitative relations between causes and effects. However, I suggest that the use of nomological machines in clinical research leads people to make a mistake analogous to the one Cartwright claims is made by those who accept fundamental laws in physics as expressing the 'real' behaviour of objects. The behavior of a body influenced only by the gravitational force between it and one other body is not the real behaviour of that body, or at least no more real than any other behaviour of that body; it is just the way that that body behaves in the context of the nomological machine that shields it from other influences besides the force of gravity. Similarly, the estimate of the effects of a drug found in a clinical trial does not reflect the 'real' effect of that drug, but its effect in the context of the trial. Both Cartwright's fundamentalists and the proponents of EBM are making the error of assuming that the development of a successful nomological machine means that the real effect has been isolated. While this does not matter so much in physics (Cartwright's point is metaphysical, not methodological), where false laws can be used to good effect in many cases, in medicine, the results have the potential to be disastrous.

Recall that a capacity is simply the ability possessed by an object to play a certain causal role in some circumstances. Cartwright notes that the capacities studied in physics have 'an exact functional form and a precise strength, which are recorded in its own special law' (p. 54). The Coulomb capacity, for example, 'describes the capacity that a body has *qua* charged' (p. 53). Moreover, she continues, we also know explicit rules for combining the Coulomb capacity with other capacities described by different force laws, thus enabling us to predict, to a reasonable degree of accuracy, how a body will behave when subject to different types of force. Conversely, we can sometimes abstract the force due to the Coulomb capacity from the total force between two particles. (Again, all of these relationships and capacities are expressed in mathematical terms.)

The capacities in which biomedical researchers are interested, however, can generally not be characterized in precise mathematical terms, though their effects can, of course, be measured, often in a variety of ways. In a clinical trial, for example, we are interested in the capacity of a drug to treat a disease or its symptoms and in its capacity to cause harm. All of these effects are quantified and compared with the capacity of (for example) a placebo to do the same things. However, we can seldom express the capacities of the drug using equations, nor can we characterize mathematically how combining the drug with other factors (changing the nomological machine) will affect the expression of its capacity to heal or to harm. In physics, the results of combining different capacities can often be predicted on the basis of abstract, theoretical laws. In biology, they cannot. We need, instead, to tell a causal story to explain how the capacity is expressed differently in different situations (for example, that antimycotic medications are less effective in treating onychomycosis in diabetics because these patients have poor peripheral circulation). Again, there is generally no way to predict these manifestations of a drug's capacities in many of the patients seen in clinical practice solely from the way in which these capacities are manifested in the controlled environment of a clinical trial. However, there is a strong tendency among the proponents of EBM to take the effects of a medication observed in a clinical trial for the real effects of the drug, or at least a good estimate of the real effects. In fact, a central aspect of EBM is the development of 'secondary resources' that review the literature bearing on a specific medical question. A subset of these studies, meta-analyses, statistically combine the results of several RCTs in order to provide a more precise analysis of this 'real' effect. What meta-analysis amounts to in my terms, however, is simply describing the results of running the same nomological machine over again.

11.7. META-ANALYSIS: RE-RUNNING THE NOMOLOGICAL MACHINE

The motivation for conducting meta-analyses is prima facie reasonable. If one clinical trial provides evidence for the efficacy of a treatment, then a number of them should provide better evidence, either confirming the conclusions drawn from the initial trial (in cases where subsequent trials have similar results to the first) or prompting us to modify our initial assessment (where the results of different trials are in conflict). Like a single RCT, however, systematic reviews and meta-analyses on their own may result in an oversimplified view of the nature and extent of the 'evidence' supporting a course of action.

Reviews of the medical literature have been in existence for decades, perhaps for as long as there have been medical journals. The traditional review article provides a summary and critique of the literature in a particular area and serves both to orient newcomers to a field and to help seasoned practitioners to keep up with recent developments. A significant contribution of EBM has been to challenge and revise the accepted format of these reviews. Traditional reviews (now described as 'narrative reviews') are varied in both their approach and their quality; some may be highly idiosyncratic, perhaps published on the strength of the author's reputation (blind peer review is uncommon in medical journals). Recently, efforts have been made to structure and standardize the methodology of these review articles. These 'systematic reviews' are themselves scientific endeavours, setting forth a question to be answered in advance of the analysis and making explicit their methods for gathering and analysing data. A subclass of systematic reviews, meta-analyses, combine the data from a number of studies quantitatively, while other reviews summarize the available evidence qualitatively. In the following, I will focus on the rationale of the meta-analysis; similar criticisms could be leveled at the qualitative review to the extent that it shares the goal of a meta-analysis to provide a more accurate assessment of a drug's effects.

A meta-analysis, then, is a type of systematic review that uses statistical techniques to combine the results of a number of trials of the same or a similar treatment regimen.[5] For any single clinical trial, the value obtained for an endpoint of interest in the study sample may vary purely by chance from the value that would be found in the population as a whole. A 'confidence interval' can be calculated that uses the actual value found in the study, together with other information such as the size of the study sample, to calculate to an arbitrary probability, often 95%, the upper and lower values between which the value actually lies in the population. The wider the confidence interval (i.e. the broader the range of possible values in the population as a whole), the less helpful the estimate of the value obtained from the sample of individuals measured in the trial will be in characterizing the value of the endpoint that would be found in the population. It is important to note that it is the *mean* value in both the sample and the population that is being discussed. By combining the results of different studies (assumed to be on the same population and measuring the same endpoint), the meta-analysis can 'shrink' the confidence interval around the point estimate of the efficacy of the study drug, thus providing a more accurate estimate of the value that would be found in the entire population.

Recall the citation from Wulff et al. earlier in the chapter, which suggested that, in a trial with a 95% confidence interval of 15–45%, a clinician would be justified in concluding that between 15 and 45% of her patients will benefit from the drug. A meta-analysis of a number of trials might shrink that range to, for example, 27–33%, giving the physician a better idea of how many patients might benefit from the drug.

However, the interpretation of the meta-analysis poses the same problems to the clinical as did the original, single, clinical trial. She still needs to determine which patients might be in that 27–33%. Yet, since the inclusion and exclusion criteria for trials of the same drug are likely to be similar (if not standard), the problem of extrapolation to patients who would not have qualified for the trials still remains. Moreover, the problem of the variability of response in patients similar to the trial patients remains and is even exacerbated. The results of a clinical trial, as previously noted, are reported as averages; the meta-analysis basically takes the average of those averages (resulting in the narrowing of the confidence interval), but it does so at the price of reducing information about the variability of responses to the study drug.[6]

In addition to these statistical issues, there are practical problems plaguing the meta-analyses and other systematic reviews currently performed. Ken Goodman (2003) emphasizes that a meta-analysis can only be as good as its raw material; thus the quality of the individual studies must be assessed prior to pooling their results. Moreover, while recent efforts to establish and enforce the registration of clinical trials with a central body may mitigate this problem, often those who conduct a review are limited to including only those trials that that have been published. Generally, though only those trials with a clear (usually a clear positive) result tend to be published, creating a 'publication bias' that hampers attempts to conduct reviews. And while historically, published trials have also generally been those that attract outside funding, recent efforts on the part of the International Committee of Medical Journal Editors (ICMJE) to limit the influence of (potentially biased) pharmaceutical industry sponsors on the results of clinical trials, together with the increasing tendency of these corporations to enlist the help of non-academic clinicians and of contract research organizations (CROs) to conduct their research means that a number of in-house trials (conducted to satisfy regulatory requirements) may never make it into print. There is also the 'file drawer' phenomenon, in which scientists themselves suppress the results of negative or inconclusive studies (Goodman notes that this is likely due to investigators' knowledge of publication bias). All of these phenomena also have the effect of making the trials included in a meta-analysis more homogeneous, which makes the estimate of a drug's effects appear to be more accurate (through shrinking the confidence interval around it).

Yet it is with the notion of the 'accuracy' of the estimate that I am primarily concerned. I suggest that this is another manifestation of the effects of mistaking the results of the operation of a nomological machine for the 'real' phenomenon of interest. In her discussion of nomological machines, Cartwright notes that part of the job of a nomological machine is to give repeatable results. However, there are two senses in which the term 'repeatable' can be understood. The first sense in which an experiment is repeatable is that 'if it were rerun in the same way with the same apparatus, it should generate the same behavior' (1999, p. 83). A good clinical trial is repeatable in this sense: running a trial with the same protocol (or one that is relevantly similar) should produce results that are the same (within certain confidence limits) as the original trial. The meta-analysis shows that this is indeed the case; on the assumption that the differences in the value of an endpoint measured in different studies are due to chance or to error, the smaller confidence interval around the value obtained in the meta-analysis than in any of the trials is evidence that the repeated trials are measuring the same thing. Thus, as Cartwright notes, an experiment that is repeatable in this sense gives as a result a 'general (albeit low level) law' (ibid.).

However, Cartwright's nomological machines also give results that are 'repeatable' in a second sense, in that they can lead to higher-level generalizations: 'the results of an experiment should be repeatable in the sense that the high-level principles inferred from a particular experiment should be borne out in different experiments of different kinds' (ibid., pp. 89–90). On Cartwright's analysis, it is this sense of repeatability that is necessary if we are to claim that we understand something about the nature, or the capacities, of the feature under study. To return to Cartwright's earlier example, we might learn more about the nature of the Coulomb capacity by studying its manifestation in different situations. The generalizations that can be made about this capacity depend on understanding the reasons for the similarities and the differences in its manifestations in different circumstances.

Neither RCTs nor the meta-analyses that combine them are designed to be repeatable in this sense. As discussed above, RCTs tend to report average outcomes and give little analysis of the variability that may occur in results between different trial patients.[7] This tendency is even stronger in meta-analyses; the narrower confidence interval reported in meta-analyses is due in part to the loss of within-groups variance that occurs when these averages are averaged. Thus, the point of a meta-analysis is to give a more precise estimate of an average effect in a population rather than to draw conclusions about the relationships between differences in the results of different RCTs and differences in the way they were conducted. It is as if it is tacitly assumed that the average effect in the population is the 'real' effect that the various trials (and the meta-analysis) aim to find and that the variance within groups is truly error, rather than real differences in the manifestations of the drug's capacity. While this approach may be fine for some purposes, it is far from clear that it is a good approach for informing clinical decision-making. What would be required instead is a scientific approach that reflects Cartwright's second sense of repeatability; clinicians will want to know what factors might affect the way in which a drug's capacity is exercised and which of those factors are relevant for an individual patient's care. The meta-analysis instead offers them the effects of a drug in a hypothetical 'average' patient.

Thus in both the RCT and the meta-analysis of a number of RCTs, the goal is to provide a precise estimate of the effects of a drug, including its intended effects and also potentially harmful side effects. While these goals are important, I have argued in this paper that they are not goals that allow clinicians to make treatment decisions for those patients who are not relevantly similar to the study subjects (for example, for patients with comorbid conditions or those taking other medications that may interact with the study drug). More subtly, but equally important, RCTs do not allow conclusions to be drawn about the variability of the response of patients *in* the study – and in almost any study, some patients will fail to benefit from (or even be harmed by) a treatment that is overall beneficial. By characterizing RCTs as nomological machines, I hope to have shown that while they may be a necessary contribution to medical decision-making, they can never be sufficient. What is required as well is research that teases out the capacities of the study drug in different groups of patients, allowing physicians to make more accurate predictions about the response of a particular individual to the treatment being considered. I acknowledge that conducting and interpreting such research will be difficult, and that even the best evidence cannot provide certainty in clinical medicine. However, it is necessary to move beyond the current focus on clinical trials to improve the 'evidence' used in evidence-based medicine.

Department of Philosophy, University of Western Ontario, Ontario, Canada

NOTES

[1]   This assumption pertains to random sampling from the population as well as random allocation of the sample to groups. Since participants in clinical trials are never randomly sampled from the pool of all eligible patients (trial samples are 'convenience samples') the assumption of randomization is always violated.

[2]   Cartwright does not here emphasize the notorious difficulty faced by such empiricists in cashing out the notion of 'necessity.'

[3]   'Real' here should be understood in the sense of 'basic' in general, regularities that are produced by a nomological machine are taken to be basic laws that underlie all regularities occurring in different circumstances. Cartwright argues against this view in the context of physics and I extend her discussion of nomological machines and capacities to clinical research.

[4]   The *NNT* is equal to one over the absolute risk reduction (i.e. the difference in risk of an adverse outcome between the experimental and the control groups).

[5]   The degree of similarity required is a vexed question in the field.

[6]   Some people advocate conducting meta-analyses on the basis of individual patient data, rather than of group data, in these studies the patients from each of the original studies are viewed as one large group, rather than a number of smaller groups. As a result, the loss of variability in results is minimized. Individual patient data is not, however, always available to those conducting a meta-analysis.

[7]   In statistical analysis, variance within a study group is known as 'error' – this terminology obscures the fact that there may be good reasons for that variance that we might wish to study.

REFERENCES

Cartwright N (1983) How the laws of physics lie. Oxford University Press, New York

Cartwright N (1999) The dappled world: a study of the boundaries of science. Cambridge University Press, Cambridge MA

Clarke M, Chalmers I (1998) Discussion sections in reports of controlled trials published in general medical journals: Islands in search of continents? JAMA 280:280–282

Crawford F, Young P, Godfrey C, Bell-Syer SE, Hart R, Brunt E, Russell I (2002) Oral treatments for toenail onychomycosis: A systematic review. Arch Dermatol. 138(6):811–816

Dennett D (1995) Darwin's dangerous idea: Evolution and the meanings of life. Simon and Schuster, New York

Goodman K (2003) Ethics and evidence-based medicine: Fallibility and responsibility in clinical science. Cambridge University Press, Cambridge MA

Goodman SN (1999) Toward evidence-based medical statistics. 1: The *P* value fallacy. Ann Intern Med. 130(12):995–1004

Gupta AK, Kohli Y (2003) Evidence of *in vitro* resistance in patients with onychomycosis who fail antifungal therapy. Dermatology 207(4):375–80

Guyatt G, Rennie D (2001) The users' guide to the medical literature. American Medical Association, Chicago

Jadad A (1998) Randomized controlled trials. BMJ Books, London

Moyé L (2000) Statistical reasoning in medicine: The intuitive *p*−value primer. Springer-Verlag, New York

Sackett DL, Rosenberg WM, Gray JA, Haynes RB, Richardson WS (1996) Evidence-based medicine: What it is and what it isn't. BMJ 312:371–72

Wulff HR, Andur-Pederson S, Rosenberg R (1990) Philosophy of medicine: An introduction, 2nd eednn. Blackwell Scientific Publications, London UK

MIRIAM SOLOMON

12. THE SOCIAL EPISTEMOLOGY OF NIH CONSENSUS CONFERENCES

12.1. INTRODUCTION: A BASIC DESCRIPTION OF THE CONSENSUS DEVELOPMENT PROGRAM

The NIH Consensus Development Program, which has been a model for national and international consensus conferences in medicine, has operated since 1977, and has produced (as of the time of writing) 118 Consensus Statements and 26 State-of-the-Science Statements. At this point, it is a social epistemic institution with a history of (some) change and adaptation to criticism and changing circumstances. It is a rich topic for social epistemic investigation.

The idea for the Consensus Development Program started in the 1970s with a request from Congress to ensure that new medical technologies, often developed at NIH at taxpayer expense, are put into appropriate use (Ferguson, 1993, pp. 180–198). The thinking was that these technologies need both unbiased evaluation and effective dissemination. The NIH Office of Medical Applications of Research (OMAR) was established in 1978 to handle the Consensus Development Program.

The original epistemic model was that of 'Science Court': a 1960s idea of Arthur Kantrowitz (see Ferguson 1997 who cites Kantrowitz, 1967, pp. 763–64 & Kantrowitz, 1976, pp. 653–56). Kantrowitz appreciated that sometimes decisions (such as funding or technological development) need to be made before a scientific issue is definitively settled. The idea is that a panel of 'sophisticated scientific judges', who are experts in scientific areas other than the one under discussion, listen to the evidence from experts on both sides of a controversy and come to an impartial, if only temporary, conclusion about the scientific issue. This, Kantrowitz thought, will keep the political issues out of the scientific arena. Kantrowitz had in mind that 'science court' would convene over controversial matters such as, 'Do fluorocarbons damage the ozone layer?' and 'How dangerous are nuclear power plants?' The scientific matters thus 'settled,' questions of policy and societal value are left to the usual political process. A general 'science court' was never implemented, but the idea was adapted for the NIH consensus program.

The basic design of a NIH Consensus Conference is as follows. OMAR, a NIH Institute or Center, another Government health agency, Congress, or the public suggests topics. OMAR decides which topics to pursue. A planning committee (composed of federal employees from the OMAR, the relevant Institute or Center at NIH, outside experts, patient advocates and the pre-selected panel chair) convenes to decide the questions that will frame the conference. Panel members for the consensus conference are chosen from clinicians, researchers, methodologists and

167

H. Kincaid and J. McKitrick (eds.), Establishing Medical Reality, 167–177.
© 2007 *Springer.*

the general public. Federal employees are not eligible as panel members, so as to avoid the appearance of government influence. Since the late 1980s, those chosen as panel members are disqualified if their own research or stated opinions can be used to answer the questions under debate (this is termed 'intellectual bias'). Since 1993, prospective members are disqualified if they have a financial conflict of interest (it has become increasingly difficult to find panelists without financial conflicts of interest). There is also no sponsorship of conferences by private corporations. (This unfortunately cannot be said about non-NIH consensus programs.) The general idea is that the panel should be seen as a neutral judge and jury. Speakers are requested, but not required, to disclose financial conflicts of interest. Speakers are not expected to be without 'intellectual bias,' but they are asked to present as researchers rather than as advocates. Members of the planning committee may also be speakers. Obviously, members of the panel are never speakers.

Consensus conferences have always been open to the public and are now broadcast on WebCam. Panel members and audience listen to academic presentations from 20–30 experts on the debated issues, and have an opportunity to ask questions. The ratio of presentation to discussion is about 2:1. The panel begins to draft a consensus statement on the first evening of the conference, and continues on the afternoon of the second day, often working well into the night. On the morning of the third day, the draft consensus document is read aloud to all attendees, and comments and discussion are welcomed. The panel recesses for about two hours to incorporate any changes, and then a press conference is held releasing the major conclusions of the consensus conference. The consensus statement goes through one more round of revisions from panel members and the panel chair before publication about a month later. Consensus statements are released to medical practitioners, policy experts and the public.

According to program guidelines, the important epistemic criteria for selection of a medical topic for the consensus program are: (1) the consensus can be evidence-based (that is, there is already enough evidence to come to a consensus: contrast this with the Science Court idea of Kantrowitz in which there is not yet strong enough evidence for consensus) (2) the matter is controversial and (3) there is a gap between theory and practice (Ferguson, 1995, pp. 332–336 & consensus.nih.gov). (These criteria are probably not all necessary for each conference. Different NIH texts highlight different criteria.) NIH says, 'The timing of the conference should neither be so early in the development of a new consensus that the data are insufficient nor so late that the conference merely reiterates a consensus already reached by the profession (consensus.nih.gov/about/process.htm). In particular, the consensus program is expected to operate in the interval between clinical trials ('investigation') and state-of-the-art guidelines ('general use') (Ferguson, 1995). Other institutions (e.g. the Agency for Health Care Research and Quality, or AHRQ—formerly the Agency for Health Care Policy and Research, or AHCPR–and professional societies) produce the state-of-the-art guidelines. NIH hopes and intends that such guidelines will be informed by prior consensus conferences. NIH is also comfortable with the influence that consensus conferences may have on reimbursement practices.

Notwithstanding, a central purpose of the Consensus Development Program is to change medical practice at the level of the individual clinician.

12.2. EPISTEMIC CONCERNS ABOUT CONSENSUS DEVELOPMENT CONFERENCES

There are two reasons for epistemic concern about this description of consensus conferences. First, it looks as though the window for usefulness is small—after there is enough evidence to reach a conclusion but before the research community itself has reached consensus. It is a consensus *development* program. The Consensus Development Program is clear that the consensus must be 'evidence-based,' that is, that enough evidence must be there. (Kantrowitz's 'science court' was, on the contrary, designed for situations in which there is inadequate evidence but a need for action.) Does this interval of time coincide with the actual time of the consensus conference, which requires over a year of preparation and planning? And why the departure from Kantrowitz's goal, to reach a 'rational' consensus before the evidence is in? A more recent epistemic concern about the idea of consensus development is, wouldn't it be quicker, more timely, and at least as good to do a meta-analysis of the available evidence? Such a formal analysis would have a similar claim to be free from bias.

The second concern about this description of consensus conferences, of more salience to those working in science studies than to medical researchers, is the episte-mology of the 'science court' model itself. Thirty-five years ago, when Kantrowitz first published the model in *Science*, Thomas Kuhn was barely heard of and the matter of attaining scientific objectivity appeared far simpler than it does today. From the work of historians of science, sociologists of science, anthropologists of science, feminist critics, social psychologists and decision analysts, we now know much more about the variety and pervasiveness of bias.[1] No one has designed a group (or individual) scientific practice in which bias is eliminated, or even reduced to insignificant levels. The consensus program 'science court' is not designed to be free of biases such as, for example, group dynamics, ordering of speakers, rhetorical force of speakers, peer pressure, chair style, general medical practice biases (e.g. intervention is generally favored over non-intervention), unsystematic evaluation of evidence, the effects of sleep deprivation and conservativeness or radicalism of panel members.[2] The only biases it is designed to eliminate (and it may or may not succeed in doing so) are those of governmental pressures, commercial pressures and biases from one's own prior research in the area.

The idea that a group of qualified persons can rationally deliberate a contro-versial issue to the point of rational consensus is powerful both politically and intellectually. Many of our social institutions and practices are regulated by such procedures and ideals (e.g. governing boards of public and private institutions, Federal investigative and advisory panels, study sections, even the idea of John Rawls' 'original position'). Indeed, the endless series of committee meetings that many of us suffer through are designed with the ideal of rational social deliberation. Their purpose is not simply political—to ensure the appearance or reality of democracy (for that, a mail ballot would be enough!) or to forge *any* agreement for the purposes of action—but also epistemic, to ensure the best (most effective, or most 'rational') decision. The point of this paper is to take a close look at one of these social deliberative epistemic procedures—NIH Consensus Development Conferences—and see if the process really yields the desired product.

12.3. EPISTEMIC ASSESSMENT

Both epistemic concerns have developed in practice. Consensus conferences seem to miss the intended window of epistemic opportunity: they typically take place *after* the experts reach consensus. For example, the 1994 Consensus Development Conference, 'Helicobacter Pylori in Peptic Ulcer Disease,' took place after the important clinical trials (in the late 1980s and early 1990s, some sponsored by NIH) and after research scientists, and many prominent clinicians, had reached consensus on the use of antibiotics for peptic ulcers. The 2002 Consensus Development Conference, 'Management of Hepatitis C: 2002' repeats recommendations that were already stated by the FDA in the previous year. John Ferguson, a longtime director of the program, wrote in 1993, 'Often the planners of any given consensus conference are aware of a likely outcome and use the conference as a mechanism to inform the health care community'. This acknowledgment displays an official lack of distress about the fact that consensus conferences typically take place too late to bring about consensus. The dissemination of knowledge goal—'closing the gap between theory and practice'—appears to suffice as a justification for going through the process. (Possibly, this is *ex post facto* rationalization.)

Consensus conferences also tend to fail to produce consensus when consensus does not already exist. NIH sometimes tries to pre-empt this outcome by designating the conference a 'State-of-the-Science Conference' (before 1999 called 'Technology Assessment Statements') in which consensus is not the goal. On other occasions, it releases dissenting statements along with the consensus statement (e.g. the 1997 'Breast Cancer Screening for Women Ages 40–49) just as is done in Supreme Court Justice cases (the NIH's analogy, not mine). This has happened only three times.

Consensus conferences have often been accused of bias, or at least of a less-than-objective assessment of the available evidence. They have been evaluated on a number of occasions: an internal review in 1980, a Rand Corporation review in 1983, a University of Michigan study in 1987, an IOM study in 1990, and most recently by an NIH Working Group in 1999. Concerns have been regularly expressed about panel selection to ensure 'balance and objectivity,' speaker selection that represents the range of work on the topic, representation of patient perspectives and more careful and systematic assessment of the quality and quantity of scientific evidence. Concerns have also been expressed about the time pressure to produce a statement in two-and-a-half days, and especially the lack of time for reflection or gathering of further information. (www.nih.gov/about/director/060399a.htm) Such concerns have in fact been behind changes in the NIH Consensus Development Program, and also behind the creation of different procedures at other kinds of consensus conferences, on both the national and international scene.

12.4. CHANGES TO THE NIH CONSENSUS DEVELOPMENT PROGRAM

Over the twenty-eight years that the Consensus Development Program has been operating, there have been some notable changes in procedure. The most significant ones are increased public participation and dissemination (this reflects the general trend towards increased public knowledge and participation in medical decision making), requirements for panel members to be perceived as without bias, more effective dissemination of results (e.g. as journal articles—primarily *JAMA*, information tables at medical conferences, direct mailings, and WWW postings) and, most recently, prior meta-analysis of evidence by AHRQ (The Agency for Healthcare Research and Quality). More of this later.

Some of the suggestions for improving the objectivity of the process have been resisted. For example, the 1999 Working Group recommended the formation of an advisory group to 'review and approve topics, panel membership and planned speakers to assure balance and objectivity'. OMAR has not done this, perhaps because it would add one more layer of bureaucracy to a process that already has oversight from a planning committee, OMAR, and open public participation. The Consensus Development Program has also resisted suggestions to remove the pressure of late night writing sessions by lengthening the conference or adding additional meetings. Perhaps it does not want to give up the motivating force and rhetorical power of a press conference at the conclusion of the meeting.

The Consensus Development Program has, however, responded to many critical suggestions with changes in procedure. In general, the changes implemented show that the Consensus Development Program has 'kept up with the times' by incorporating important trends in health care research and patient involvement. Over the years, the qualifications for 'freedom from bias' for panel members have been strengthened and include both intellectual and financial conflicts of interest. Patient advocates have been included in the planning process. And, very recently (since 2001) and notably, the Consensus Development Program has created a partnership with AHRQ, which produces a systematic review of the evidence, made available to the panel in a one-day executive meeting about a month prior to the Consensus Conference, which allows opportunity for reflection and follow-up. These preparations are (in the words of Susan Rossi, Deputy Director of OMAR—telephone conversation 9/23/02) 'about a year of work'. Meta-analysis of evidence has gained widespread use in medicine over the past ten years, and the consensus program is wisely making use of this technique. The 'evidence-based medicine' approach, which makes use of meta-analysis, has rapidly become the standard of knowledge in medicine.

12.5. NEW EPISTEMIC CONCERNS

Use of the AHRQ to provide meta-analysis of the evidence only adds to the epistemic concerns mentioned above. What is there left to do—that is also epistemically respectable—after the AHRQ has done its work? Why not simply disseminate the AHRQ review? Why bother with the whole public show of expert speakers and 'neutral' jury? Since the consensus development program is committed to producing 'evidence-based consensus' and the AHRQ supplies the panelists with a systematic review of the evidence, it looks like there is nothing of scientific value left for the panel to do. When I asked this question of Susan Rossi, I was told that the panel can still do 'interpolation' or 'extrapolation' (telephone conversation 9/23/02). What is meant by this (according to Rossi) is not 'extrapolation of research results'— which would be epistemically suspect—but the first steps towards applying the scientific results in the health care arena, for example, a 'comparison of risks and benefits'. Interestingly, the NIH consensus program was initially conceived to leave out discussions of application as far as possible, and simply to evaluate the state of research. (Doesn't it look as if the Consensus Development Program is finding products to justify its continued existence? More of this later.)

The NIH consensus development conferences are also a victim of their own success. It was a new idea in the 1970s to bring together a balanced group of experts to produce health recommendations. But now, every self-respecting country and professional organization is doing it or something like it: guidelines, consensus statements and policy statements abound.

Moreover, many of them (designed after the evidence based medicine movement took hold) rest on systematic reviews of the evidence. Clearly, NIH had to incorporate systematic review of the evidence or lose its credibility. Does the NIH consensus program offer anything special or anything superior in this newest scene? Or are bureaucratic inertia or personal interests inside OMAR maintaining the program?

12.6. RECENT CASES: NIH CONSENSUS STATEMENTS SINCE INSTITUTION OF AHRQ EVIDENCE REPORT

There has been a break in proceedings, and then four consensus conferences since the addition of the AHRQ evidence report to the preparation for the conferences: 'Diagnosis and Management of Dental Caries Throughout Life' (2001), 'Management of Hepatitis C: 2002' (2002), 'Total Knee Replacement' (2003) and 'Celiac Disease' (2004).

I have taken a look at the first two, and found contrary and baffling results. The consensus statement on Hepatitis C echoed the AHRQ findings on diagnosis of infection, diagnosis of liver damage, and effectiveness of available treatments. It said more only about practical recommendations (e.g. needle exchange programs) and research recommendations (e.g. proposes a Hepatitis Clinical Research Network). This is the kind of result that Rossi was suggesting, and it does look as though the panel did no science, but simply reflected a consensus that already exists among experts on Hepatitis C. (The 2002 consensus development program statement on Hepatitis C also does not significantly differ from the FDA statement of the previous year.)

The consensus statement on dental caries, however, appears to reject the standard of evidence used by AHRQ: while AHRQ found little evidence for any of the diagnostic, preventative and treatment modalities under study, the consensus development program reported significant positive evidence for a number of preventative techniques (e.g. use of sealants, fluoride gels, fluoride varnish, chlorhexidine gels, and sorbitol and xylitol as sweeteners) and treatments for early decay (e.g. fluoride in water, chlorhexedene varnishes and gels, fluoride toothpaste and varnishes, sealants) that are already in widespread use. The consensus statement does acknowledge that little research in dentistry is up to recent standards of evidence-based medicine, and urges that future research meet the higher standards.

I hazard some guesses as to what happened here. Most of the panelists and experts in the consensus conference are dental professionals with salient experience and vested interests in the general standard of care. Methodologists who are experts in assessing the quality of evidence, on the other hand, did the AHRQ report. Dental practice—like most medical practice—is based on an evolved set of practices, few of which have ever been systematically tested. The consensus development program speakers and panel probably thought that AHRQ was throwing out the baby with the bathwater, and implicitly (interestingly, not explicitly) lowered the standard of evidence so that more current practices would appear justified. They do not dare to explicitly disagree with the AHRQ report or its standards of good evidence; indeed, they urge compliance with these standards in the future.

12.7. DISCUSSION

The above considerations lead me to conclude that NIH consensus development conferences, whatever they do, do not bring about rational consensus on controversial health topics. Usually, a consensus exists beforehand, at least among the researchers. These days, the rational basis for that consensus is made clear by the AHRQ formal assessment of the evidence. If the AHRQ assessment shows that evidence is lacking for a consensus, then consensus is not dissolved either: as we saw in the case of the conference on dental caries. What, then, if anything, do NIH consensus conferences accomplish?

NIH has stated that it sees the consensus development program as having a dual mission: evidence-based consensus formation and closure of 'the gap between current knowledge and current practice.' The second part of the mission, sometimes called 'knowledge transfer' (http://consensus.nih.gov/about/process.htm), may have been an *ex post facto* 'add on' to the original mission, but it is a worthy project and it deserves evaluation on its own terms. The NIH makes great efforts to release the consensus statements to practitioners and to the public. It has increased these efforts (with mass mailings, web site availability, publication in professional journals etc) after evaluation showed that the consensus statements had little effect on practice.

Closing 'the gap between current knowledge and current practice' is difficult. Many health care practitioners are out-of-touch with research findings, slow to change their practices, or unable/unwilling to change. Information from a credible source is likely to be part of the machinery of change.

The NIH consensus conference program—at least in its ideals–fulfills most people's conception of what a 'credible source' is. NIH is the most prestigious and largest health care research facility in the world. The Office of the Director (in which OMAR is situated) arranges consensus conferences, and panelists are independent of commercial, governmental or prior research interests. The panel includes at least one public representative, and all sessions (except for the sessions where the panel deliberates and decides, like a jury) are open to the public. The conferences do not accept financial support from professional groups, pharmaceutical companies or other commercial interests. This distinguishes NIH consensus conferences from most other similar conferences, which rely on support from professional organizations (e.g. the Eastern Association for the Surgery of Trauma sponsored conferences in the mid 1990s for establishing guidelines) that may have their own professional biases, or support from commercial interests (e.g. Canadian consensus conferences, which permit support from pharmaceutical companies). The NIH consensus program works hard to be perceived as 'objective'. While it may not actually achieve objectivity *qua* freedom from all bias (some doubts about this were expressed above) it goes a long way towards capturing what most people think 'objectivity' is, namely, the 'freedom from conflicts of interest'.

The NIH consensus program has never been assessed for the accuracy of outcomes. No-one has investigated, for example, whether the outcomes are better—more 'true' or whatever—than those achieved by other methods such as non-neutral panels or formal meta-analysis of evidence. What the NIH consensus program has going for it is its design to eliminate what most scientists and physicians think are the most troublesome forms of bias. After all, most have not heard of Thomas Kuhn, let alone Amos Tversky and Daniel Kahnemann.

Since a goal of the NIH consensus conference is dissemination of new research in order to 'close the gap between current knowledge and current practice,' then the *appearance* of objectivity is more important than a reality of freedom from bias. Scientific knowledge is communicated on trust, and trust is all about the *perception* of reliability. Trust, in our culture, is transmitted primarily through concrete human relationships (see e.g. Steven Shapin 1994). Health care practitioners may neither trust nor understand the results of a statistical meta-analysis, even if done by AHRQ. If health care practitioners trust the results of NIH consensus conferences, which have a social epistemology that *seems* trustworthy, then they will be more likely to believe them and put them into practice. Of course, if the results of consensus conferences are, in fact, shown to be frequently unreliable, that will damage any perceptions of reliability and objectivity. Successful dissemination of the results of consensus conferences depends both on the consensus conferences achieving an epistemically accurate consensus and on the consensus conferences maintaining the *appearance* of objectivity. Thus, prior consensus by experts (and by meta-analysis of evidence) is a *plus* rather than a disqualification for an effective consensus conference; a good outcome is assured by what the experts do (or what the AHRQ analysis shows) and good dissemination is assured by what the consensus conference panel does to maintain the appearance of objectivity. This leads to the conclusion that the role of consensus conferences is rhetorical (persuasive) rather than productive of knowledge.

If this sounds cynical—and it does to me (I will soften the cynicism later!)—what I have to say next is even more cynical. Despite their best efforts, NIH consensus conferences have had little direct effect on medical practice (Ferguson, 1993). Ignorance and reluctance or inabilities to change are huge forces to contend with. Where NIH conferences succeed, they do not succeed alone: they succeed because there are incentives for health care practitioners to change, such as new reimbursement policies, or rewarding campaigns by pharmaceutical companies who stand to profit from a widespread implementation of the new consensus statement. For example, the conclusions about the effectiveness of liver transplantation (1983) were more quickly implemented because they were in the interests of hospitals and surgeons. Pharmaceutical companies quickly seized on the conclusions regarding the use of steroids for prematurity in pregnancy (1994). While NIH itself avoids ruling on matters of policy, or producing particular guidelines for particular social contexts, it cannot prevent (and often does not want to prevent) others from using the consensus statement information to do this.

Is this a sorry state of affairs? The self-presentation of consensus conferences is that they produce an epistemically worthwhile consensus by the use of an objective procedure. The reality is that the real epistemic work is done beforehand, and the consensus conference is a rhetorically efficacious way to get the word out, to interested intermediaries such as professional groups, pharmaceutical companies and health insurance companies who will then adapt the statements for their own particular purposes.

There are two reasons to soften this cynical conclusion. First, it is better to have the policy, guidelines and reimbursement rules that motivate most health care practitioners be at least guided by research results, rather than by the interested parties themselves. Without the NIH consensus statements and similarly produced statements, for example, health care insurance companies might try to come up with different guidelines more suited to their bottom-line interests. Second, formal meta-analysis of research, classified into grades of quality of research, is hardly user-friendly. The NIH consensus statement is written so as to be intelligible not only to primary health care practitioners but to health care administrations and the general public. NIH consensus conferences are not only rhetorical forces; they make the research more widely accessible.

12.8. CONCLUSION

The NIH Consensus Development Program has not realized any of its official epistemic missions. As conceived by Kantrowitz, the idea was to achieve a preliminary unbiased scientific consensus on matters of public concern. As originally instituted in the late 1970s, the idea was to achieve an *evidence-based* consensus, and not to convene unless adequate evidence was present. Through most of the history of the institution, consensus statements have reflected an already existing consensus in the research community, and have been re-appropriated as educational resources for the health care community. In fact, what has mattered most is the *appearance* of objectivity (the cultural ideal, but not the epistemic reality, of group expert rational deliberation), and the associated retention of epistemic power by the scientific community (rather than e.g. insurance providers). Any decision to change the NIH Consensus Development Program should take into account these actual achievements.

ACKNOWLEDGEMENTS

This paper was presented to audiences at the University of Alabama, Georgia Tech and John Hopkins University. I thank the participants for their useful comments. In particular, I thank Harold Kincaid for particular suggestions for revision for publication.

Professor of Philosophy, Temple University, Philadelphia, Pennsylvania, USA

NOTES

[1] Thomas Kuhn inspired several traditions in sociology of science from the Edinburgh School (e.g. Barry Barnes, David Bloor, and Steven Shapin) to the Bath School (Harry Collins, Trevor Pinch) to the ethnographic work of Bruno Latour, Karin Knorr-Cetina and others. All look at how social factors influence the content and process of scientific thinking. Feminist critics of science such as Sandra Harding, Evelyn Fox Keller and Donna Haraway have given examples of ideological bias in a range of empirical sciences. Solomon Asch's experiments in the 1950s and Stanley Milgram's in the 1960s showed the strong influence of peer pressure on decision making. Finally, the work since the 1970s of Amos Tversky, Daniel Kahneman, Richard Nisbett and others have shown the effects of cognitive biases such as anchoring and availability.

[2] These are not just theoretical concerns. The NIH Consensus Development Program has been evaluated on a number of occasions and these sorts of concerns raised. See the discussion in a couple of pages, below.

REFERENCES

Agency for Healthcare Research and Quality. Includes recent evidence reports. [On-line]. Available: http://www.ahrq.gov

Council on Health Care Technology (1990) Improving consensus development for health technology assessment: an international perspective. National Academies Press, Washington, DC

Ferguson J (1993) NIH consensus conferences: dissemination and impact. Ann NY Acad Sci 703:180–198

Ferguson J (1995) The NIH consensus development program. Joint Commission Journal on Quality Improvement 21(7):332–336

Ferguson J (1997) Interpreting scientific evidence: comparing the national institutes of health consensus development program and csourts of law. The Judges' Journal, Summer 1997, pp. 21–24, 83–84

Institute of Medicine Study (1990) Consensus development at the NIH: improving the program. National Academies Press, Washington, DC

Kantrowitz A (1967) Proposal for an institution for scientific judgment. Science 156: 763–64

Kantrowitz A (1976) The science court experiment: an interim report. Science 193:653–56

National Institute of Health. Website on the Consensus Development Conference Program. [On-line]. Available: http://www.consensus.nih.gov

Shapin S (1994) A social history of truth: civility and science in seventeenth-century England. The University of Chicago Press, Chicago

13. MATERNAL AGENCY AND THE IMMUNOLOGICAL PARADOX OF PREGNANCY

13.1. INTRODUCTION

Immunology is called upon increasingly to explain and treat cases of infertility and pregnancy loss that are of unknown causal origin. While the desire to explain these so-called 'occult' cases of infertility is certainly understandable, from a scientific standpoint immunological infertility treatments are being offered to women prematurely and this is cause for concern.[1] Existing data do not yet support the use of immune testing for infertility and the risks and benefits of immune therapies are currently unknown (Kallen and Arici, 2003). Indeed, there have been reports of immunological reactions—in some cases severe—in response to immunological and other kinds of infertility treatments (Katz et al., 1992; Ben-Chetrit and Ben-Chetrit, 1994; Casoli et al., 1997; Tanaka et al., 2000).

The immunological treatment of infertility is also questionable because there has been relatively little investigation of sex differences in immune function, even though significant differences are known to exist and are clearly relevant to women undergoing such treatments (Whitacre et al., 1999; Whitacre, 2001). Instead of focusing on sex differences in immune function, reproductive immunology is most concerned with the immunology of fetal implantation and gestation. By way of illustration, a recent article about pregnancy and autoimmunity expresses concern about how to 'improve the diagnosis and treatment of infertility' and obtain 'a more detailed understanding of the immune-mediated mechanisms occurring within the reproductive organs in cases of unexplained recurrent abortion and infertility' (Geenen et al., 2002, p. 322). The goal of learning more about autoimmunity in women is not mentioned.[2] The research gap concerning immunological sex differences is further exacerbated by the fact that even research focusing on fetal implantation immunology is undertaken for reasons other than learning about pregnancy immunology. Finding new ways to improve organ transplantation and cancer treatment are two such reasons (Hunt, 1996, Preface).

The premature use of immune tests and treatments for infertility combined with the lack of understanding about sex differences in immune function makes for a very unsatisfactory situation, both scientifically and ethically. My interests in this paper, however, are the ontological commitments and values that underlie this unsatisfactory situation. The focus in reproductive immunology on fetal implantation and gestation at the expense of broader questions about immunological sex differences provides important clues in this regard. A central focus of research about the immunology of fetal implantation and gestation is maternal-fetal immunological *conflict*. I argue that this focus on conflict involves two philosophically questionable

179

H. Kincaid and J. McKitrick (eds.), Establishing Medical Reality, 179–198.
© 2007 *Springer*.

positions: an ontological commitment to the idea that mother and fetus are two clearly distinct immune entities and a value-laden assessment of how these two entities relate to one another. Neither the ontological commitment to distinct immune entities nor the focus on conflict is mandated empirically. Indeed, to characterize the maternal-fetal immunological relationship primarily in terms of conflict is to minimize or ignore neutral and beneficial relationships without justification. Note that the point being made here is not that maternal-fetal relations are or should be free of conflict. Rather, the point is simply that, as a matter of fact, conflict is emphasized at the expense of other kinds of relation in reproductive immunology.

I contend that the above-mentioned ontological commitments and value-laden views of the maternal-fetal relationship are also connected in a fundamental way to problematic views of maternal agency. It is not altogether surprising to find that problematic conceptions of maternal agency exist in immunological theory. The existence of distorted social and medical conceptions of women's agency, particularly with regard to reproduction, is well substantiated (McLeod, 2002; Martin, 2001). In keeping with this, it is widely assumed in reproductive immunology that maternal immunological agency is either absent or aggressively pathological. The possibility that maternal agency can be beneficial and constructive in pregnancy is largely ignored. Though some evidence is now available to support the existence of beneficial maternal immunological agency,[3] the phenomenon remains under-investigated and problematic assumptions about agency still require articulation and evaluation.

My main objective in this paper is to remove conceptual obstacles to the idea that women's immune systems are beneficially *active* and *constructive* in pregnancy— and to argue that these obstacles are due more to a value-laden ontology than any empirical necessity. With this in mind, I will explain in section 13.1 of this paper the 'immunological paradox of pregnancy' and show its relation to the principal immunological model of maternal-fetal relations, which I call the 'foreign fetus model.' I show that the foreign fetus model is in tension with empirical and evolutionary reasoning and therein lay the groundwork for my view that the foreign fetus model is not compelling. In section 13.2, I illustrate how the foreign fetus model distorts maternal immunological agency by examining two problematic terms—'habitual abortion' and 'fetal invasion'—used frequently in reproductive immunology. In section 13.3, I argue that problematic ideas of maternal immuno- logical agency are exacerbated by evolutionary accounts that conceptualize women as either hostile to fetuses or passive victims of fetal invasion. I also consider here, from immunological and evolutionary perspectives, the current poverty of conceptual links between pregnancy immunology, the menstrual cycle, and immuno- logical defenses against reproductive pathogens. This conceptual poverty is traceable to distorted views of maternal agency as well as to a lack of research initiative concerning sex differences in immune function.

Before I begin, I should clarify that I do not intend my arguments about the immunological maternal-fetal relationship to say anything about the social and psychological maternal-fetal relationship. It may be that a woman conceives of herself as passive in pregnancy, or feels hostile towards her fetus, or feels invaded by her fetus and sees her fetus as aggressive. I do not speak to these issues. Rather, my argument is concerned with immunological relationships and the assumptions and values that surround their conceptualization. I should also clarify two presuppositions I make in this argument. The first is that values can and do shape inquiry in immunology and I contend that even 'good' immunology contains value-laden assumptions. These assumptions should be explicated and evaluated philosophically and empirically. My goal is the former, though I take the latter into serious account. My second presupposition is that the term 'agency' can be used to describe biological activity, provided that it is not understood as the conscious, teleological human form of agency. The relation between general biological and conscious human forms of agency is at least metaphorical; and given the nature of metaphorical reasoning, it seems safe to say that social and psychological views of agency influence how and to what we attribute biological agency.

13.2. THE IMMUNOLOGICAL PARADOX OF PREGNANCY

The immunological relationship between mothers and their fetuses has long been considered paradoxical. Peter Medawar, who first proposed that the fetus was akin to foreign transplanted tissue, put the problem this way:

> how does the pregnant mother contrive to nourish within itself, for many weeks or months, a foetus that is an antigenically foreign body? The question derives its significance from the fact that the mother does not always contrive to do so; it is sometimes immunized against the antigens of its foetus, with the consequence that the foetus, or its successors in later pregnancies, is either destroyed or born with afflictions that are the more or less immediate outcome of cellular damage (Medawar, 1953, p. 324).[4]

Because the fetus is genetically distinct from the mother, it is almost the immune equivalent of an organ transplant. Given this, we should expect the mother to reject it. Generally, of course, the fetus is not rejected; and this gives rise to the paradox.

Medawar's idea that the fetus is foreign and therefore in danger of maternal attack is connected to the self-nonself discrimination theory of the immune system which emerged in the late 1940s and early 1950s. This theory holds that the primary organizing principle of the immune system is its ability to learn the difference between self and non-self structures. The reasoning is that the immune system needs to distinguish pathogens from self-structures in order to know what to respond aggressively to and what to ignore. This preoccupation with the self-nonself distinction is carried over to maternal-fetal relations. But if the immune system is organized around the recognition of and aggressive response to non-self structures, and the fetus is non-self, what protects the fetus from maternal immunological attack?

Many of the contemporary answers can be traced back to Medawar's original proposals. Medawar thought that some combination of three mechanisms likely prevented maternal immune attacks on the fetus: an anatomical barrier between mother and fetus, altered or immature fetal antigens, and maternal *immunosuppression*—or as Medawar put it, the 'immunological indolence or inertness of the mother' (Medawar, 1953, p. 327). Today, some think that the fetal trophoblast cells lining the uterine spiral arteries act as an anatomical barrier. It is also thought that fetal trophoblast cells alter or suppress the expression of cell-surface antigens that would normally identify them as foreign to the mother. And, evidence exists that the maternal immune system is suppressed, particularly at the maternal-fetal interface. But it is also widely recognized that while Medawar's three hypotheses have been useful to immunologists, they 'do not completely explain how the fetus evades the maternal immune system' (Koch and Platt, 2003, p. 95). What is retained from Medawar's account, however, is the idea that the fetus must *evade* the mother. Many contemporary immunologists think that the fetus must in some sense hide from the maternal immune system and that the maternal immune system must be suppressed.

The antagonistic view of maternal-fetal relations emerging from self-nonself discrimination theory I refer to as the 'foreign fetus model.' Underlying the foreign fetus model is that idea that maternal-fetal relations are essentially antagonistic and so must be managed through barriers, evasion, and suppression. The foreign fetus model is the principal model of maternal-fetal relations in reproductive immunology and it continues to exert substantial influence, despite a number of tensions that accompany it. There are at least two sources of these tensions within reproductive immunology, each of which is ultimately related to problematic views of maternal agency.

The first tension is generated by the claim that maternal-fetal conflicts are managed through maternal immunosuppression. This explanation makes little evolutionary sense, for pregnancy is hardly the time to have generally weakened defenses against pathogens. Given this, it is at first glance quite odd that Medawar proposes his immunosuppression hypothesis in an article about the evolution of viviparity in vertebrates. In fairness to Medawar, his struggle was to understand why fetuses are not rejected given that he knew immunological 'leakage' between mothers and fetuses occurred; he knew the barrier was imperfect. This is a difficult puzzle, even if—as I argue—it is generated by the unnecessary imposition of strict self-boundaries and self-nonself conflicts onto the maternal-fetal relationship. Underlying Medawar's view, however, is the more general thesis that mammalian fetuses are evolving greater independence from threatening, inconsistent and inadequate maternal environments. Medawar argues that the fact that ovaries are not essential to the completion of pregnancy in women or monkeys, but are to pregnancy completion in say, cows and rabbits, is indication of an evolutionary *progression* 'towards a complete endocrinological self-sufficiency of the foetus and its membranes – in short, towards the evolution of a self-maintaining system enjoying the highest possible degree of independence of its environment' (Medawar, 1953, p. 324).

Medawar's endocrinological-evolutionary analysis underscores the idea in immunology that mothers and fetuses should be kept separate to prevent harm. In the case of immunosuppression, the maternal need for pathogen defense is in conflict with another evolutionary imperative: the drive towards fetal independence.

While Medawar's concerns with respect to fetal independence now seem dated, the tension between maternal immunosuppression and the need for effective defense against pathogens continues in a variety of ways.[5]

A second tension generated for the foreign fetus model concerns the hypothesis that pregnancy is more likely to occur and result in healthy offspring when the maternal and paternal genes encoding for the histocompatibility proteins known as human leucocyte antigens (HLA) are sufficiently different (Chaouat, 1993; Roberts et al., 1996; Creus et al., 1998). For partners with very similar HLA genes, pregnancy is less likely to occur and when it does, it is more likely to end in fetal loss. It appears that maternal immune recognition of the fetus is enhanced when paternal HLA is very different from maternal HLA and that this maternal recognition of the fetus is beneficial, not harmful. Thus, instead of increasing maternal-fetal conflict, as predicted by the foreign fetus model, maternal recognition of the fetus may actually decrease such conflict. This is why immunologist Joan Hunt (1996, Preface) claims that it is *surprising* that maternal immune recognition of the difference between fetal and maternal tissues increases fertility. Tension exists, then, between the proposed benefits of maternal-paternal immunological difference and the prediction of the foreign fetus model that immunological similarity should be protective.

Perhaps part of the reason for these tensions is that, as Ashley Moffett and Y. W. Loke claim, immunologists do not treat pregnancy as an immunologically unique phenomenon (Moffett and Loke, 2004). They claim that pregnancy is forced to fit self-nonself discrimination theory and other general models of the immune system, and that it should not be. One way to resolve the two aforementioned tensions, then, is to drop the view that pregnancy should fit self-nonself discrimination theory. This would enable us to escape the idea that differences between maternal and fetal tissues are inherently antagonistic. It is possible, for example, that some of the immune activities thought to be involved in immunosuppression or fetal evasion actually have other roles, such as tissue remodeling at the maternal-fetal interface or differential immune responses specific to the location of infection in maternal, placental or fetal tissues. Such functions need not concern self-nonself discrimination and this possibility raises some interesting questions. For example, why is pregnancy made to fit models of immune function based on non-pregnancy, rather than treated as a unique biological event? Why does the idea that mother and fetus are distinct immunological selves in need of protection from each other persist, despite the tensions that accompany it?

To answer these questions from a philosophical perspective, it is important to look at the notions of selfhood and agency at play in immunology in closer detail—a task to which I now turn.

13.3. HABITUAL ABORTERS, FOREIGN INVADERS

The distortion of maternal agency in the foreign fetus model is due in part to two assumptions the model entails. The first assumption is that a mother and her fetuses are two clearly distinct and separate entities. The second assumption is that the primary defining relationship between mother and fetus is one of conflict. Part of the problem here is the type of body assumed to house the 'standard' immune system. For example, Lisa Weasel (2001) contends that the self/other dichotomy in immunology assumes a masculine autonomous self that incorporates a notion of the other as pathological. The autonomous and unary male body, taken as the standard, naturally generates a paradox when applied to pregnant women. When pregnancy is made to fit this standard, mother and fetuses must each be treated as unary selves. Taking into consideration the pervasiveness of aggression-based metaphors in immunology, it is easily surmized how these unary selves might relate when they come into contact with each other. This ontological perspective is made worse by the sort of phenomenon Emily Martin (1994b) refers to as 'immune machismo,' which she illustrates with the example of a medical resident who dismissed 'the possibility of contracting HIV infection from blood in the emergency room where he worked, [claiming] his immune system could "kick ass"' (Martin, 1994b, p. 389).

The problems concerning agency that arise from an ontology privileging the unary, autonomous self are helpfully illustrated by two terms commonly used in reproductive immunology: 'habitual aborter' and 'fetal invasion.' The term 'habitual aborter' (and the somewhat more common term 'recurrent aborter') is used in studies of recurrent pregnancy loss to refer to women who have had three or more successive fetal losses (Laskin et al., 1994; Bussen and Steck, 1997; Tanaka et al., 2000; Adachi et al., 2003). These terms are now more commonly used in immunology given that immunological explanations for 'occult' cases of pregnancy loss are increasingly invoked. In addition to diminutively referring to women with this problem as 'aborters,' the 'habitual' label suggests they have a compulsion or perhaps an inherent disposition to reject fetuses. The terminology suggests, in this context, that such women have dangerous immune systems; their pregnancy losses are evidence of an immunological hostility. Maternal agency is diminished insofar as the immune system is perceived as being out of control and whatever agency the maternal body does have is considered pathological. Moreover, the implication that these women have an inherent disposition to reject fetuses hides their partners' role. There is some evidence that male partners may have a deficiency that contributes to successive pregnancy losses, either alone or in concert with a deficiency on the part of the female (Clark, 1999a). But instead of seeing women's reproductive immunity as incompatible with certain sperm, that is, as simply an indication that perhaps some reproductive matches are better than others are, and instead of looking for a problem with sperm, the tendency is to conceptualize women's immunity as hostile to sperm.

The second term—fetal invasion—is more pervasive than the term 'habitual aborter.' Its force and substance also depends on the assumption that conflict is the most important maternal-fetal relation, though here it is the fetus (or blastocyst) that is thought antagonistic to the mother. The process of implantation, for example, is typically described as fetal 'invasion.' Here, the mother is passive as the fetus invades and destroys her tissues. In fact, the behaviour of the fetal trophoblast is often compared to invasive cancer or parasitism and theoretical models based on these comparisons exist (Govallo, 1993; Clark, Arck, Chaouat, 1999; Bubanovic and Najman, 2004). During implantation the fetal trophoblast cells that come to line the maternal spiral arteries are conceptualized as 'manning' a boundary between maternal and fetal tissues. Fetal cells 'take over' maternal arteries in order to prevent a maternal immune system attack and to keep the blood supply open. Immunologists will say things such as 'there is extensive fetal invasion into maternal tissue' (Entrican, 2002, p. 80); that 'the human female is the only mammal needing the second deep wave of trophoblast invasion' (Robillard, Dekker, Hulsey, 2002, p. 106); or that 'the fetus creates a privileged boundary for its relatively short but important invasion of the mother' (Langman and Cohn, 2000, p. 192). Note here that it is the fetus that creates the boundary, not the mother. The mother is passive as fetal trophoblast cells remodel her uterine arteries, dissolve tissue, and move themselves in. And if this is to occur in that way that it should, the mother's usual immunological defenses must either be disabled or fooled.

The view of implantation as invasion is problematic for at least three reasons, however. First, tissues are dissolved and remodeled elsewhere in the body and this is not thought of as invasion or destruction. The difference in the case of pregnancy must be the assumption that there are two distinct and antagonistic beings present. But this is problematic, both biologically and philosophically. At the very least, it must be acknowledged that there is a permeable interface between mother and fetus and that the distinction between mother and fetus develops gradually. Development, after all, is *the* key feature of pregnancy (Sumner, 1981). Immune and other cells of the fetus and mother are normally found to coexist in close proximity without problems (Bonney and Matzinger, 1997; Moffett-King, 2002; Hall, 2003). Cells from mothers enter fetuses and those from fetuses enter mothers and the exchanged cells may reside in the other for *life* (Hall, 2003; Nelson, 2001, 2002; Bianchi, 2000). There is also evidence that these cellular exchanges may confer benefits to mothers and their offspring. There is, then, no clear boundary dividing mother and fetus: cells from each populate the permeable interface between them. Immunologically, the distinction between them is vague. Mother and fetus are, in a sense, 'not one, but not two' (Karpin, 1992).

Second, even if there clearly were two beings present in pregnancy, this does not mean that implantation should be thought of as invasive. 'Invasion' suggests antagonism, conflict, and war and the use of such aggression-based metaphors in immunology has not gone unchallenged (Haraway, 1989; Martin, 1994a; Weasel, 2001). However, immunologists' use of invasion terminology does not always incorporate war metaphors; it may incorporate metaphors based on host-parasite relations or the disease process of cancer. But whether war, cancer or parasitism metaphors are used, invasion terminology suggests pathology: something is wrong and it should be stopped. In this respect, blastocysts are relevantly different from invading armies, cancers and parasites. Blastocysts usually go where they are supposed to go. From the maternal perspective, cancer and parasites do not. Moreover, whether war, parasitism or cancer metaphors underlie the language of 'invasion,' an inappropriate aggressiveness on the part of fetal trophoblast cells is suggested. 'Trophoblast migration' (Van Nieuwenhoven et al., 2003, p. 347) is a more appropriate term: it accounts for the movement of fetal cells without making evaluations based on aggression or pathology.

Finally, conceptualizing implantation as invasion also assumes the mother is passive during implantation. The most the maternal is thought to do is prevent the invasion from going too far. However, if we reconceptualize implantation as something other than invasion, space opens in which to consider that the maternal may actually be positively involved in the process of implantation. Recent evidence supports such an interpretation. Some argue that the maternal immune system is *actively* involved in implantation and the construction of the maternal-placental interface (Entrican, 2002, Moffett-King, 2002). Maternal immune cells, including specialized uterine natural killer cells and $\gamma\delta$ T lymphocytes appear to actively remodel tissue. Moreover, maternal immune chemicals are thought important for the growth and development of the placenta and fetus (Hunt and Soares, 1996, p. 10; Adachi et al., 2003, p. 180; Van Nieuwenhoven et al., 2003). In short, it appears that the maternal immune system recognizes and responds to the fetus and this immune response is *beneficial*, not harmful, to pregnancy. The difficulty with fetal antagonism is that it does not accurately describe the implantation process, in part because the mother's participation during implantation is ignored. Given maternal participation, the term 'placental construction' is another plausible candidate to replace invasion terminology.

That such plausible candidates for implantation terminology have not been popular has much to do with the fact that self-nonself discrimination theory has been so influential in immunology. But the self-nonself discrimination theory of the immune system is increasingly criticized by immunologists and controversy regarding its significance continues (Langman, 2000; Tauber and Podolsky, 1997). Immunologist Polly Matzinger (1994), for example, argues that the immune system is not principally concerned with self-nonself discrimination. Rather, the system bases it decision about what to attack and what to ignore on danger—that is, cellular distress. On this view, provided that a pregnancy is healthy and there are no distress signals, the maternal immune system will not harm the placenta or fetus. Thus, there is no need to maintain a strict boundary between the mother and fetus. The 'foreignness' of the fetus is irrelevant.

One of the key difficulties in viewing maternal immune systems as primarily passive or pathological is that it masks other possible immunological relationships, from the neutral to the beneficial and cooperative. The danger theory takes a positive step forward insofar as it allows for maternal-fetal neutrality. It also suggests that some functions once thought to involve self-nonself discrimination between mother and fetus may not, in fact, be 'immunological' at all. For example, in pregnancy, some activities thought to be immunological may actually be physiological functions geared towards the maintenance of adequate blood flow between mother and fetus (Bonney and Matzinger, 1997). However, the view that beneficial immune communication takes place between mothers and their fetuses goes beyond what is proposed in the danger model.

Investigation of the ontological difference between the unary selves of the foreign fetus model and the selves of a 'not one, but not two' model may encourage the development of models of maternal immunological agency that include beneficial maternal activities. The concept of the unary self underlying the foreign fetus model often assumes an individuality that is abstract in its purity—an idealized individuality that is completely independent of and isolated from others. Concepts of agency that incorporate an idealized individuality will naturally share in this abstractness. And, if this idealized individuality is considered a prerequisite for full agency, those with unary selves that are less clearly distinguished from others may not be regarded as full agents. But there is no demand that we view selves or agency in an abstract and idealized manner. Agency need not depend upon *isolated* independence. Rather, agency can arise from an independence that is fully contextualized by its relationships to the wider world (Sherwin, 1998; MacKenzie and Stoljar, 2000; McLeod, 2002).[6] A robust form of agency is possible in a 'not one, but not two' relationship.

This more socialized concept of agency is akin to that supported by biologist Lynda Birke, who argues for

an understanding of agency that emerges out of the engagement of the organism with its surroundings; it is thus an agency in relation, not an essential property of the individual. In that sense, the 'agency' of the tissues of the developing foetus can be understood only in relation to the foetus's engagement with its environment – which must include the mother's body (Birke, 1999, p. 152).

For Birke, traditional notions of agency, with their strong emphasis on disconnected individuality, are unsatisfactory. However, so are views that treat agency as fragmented, fractured or dissolved. The goal is a robust understanding of agency that avoids the ontologically and empirically problematic isolated agent. Socialized agency, then, can be understood as an improved heuristic for understanding biological organisms in relation. This revised understanding of agency functions better empirically, especially in the case of maternal-fetal immunology: it neither requires the reification of artificial conceptual boundaries, nor ignores the complexity of maternal-fetal interactions.

But introducing a socialized form of agency to correct problematic assumptions about maternal agency will only go so far towards expanding immunology's purview. Problematic notions of women's agency also have deep roots in *evolutionary* accounts of maternal-fetal relations and these evolutionary accounts are increasingly invoked in reproductive immunology. While greater contact between reproductive immunology and evolutionary theory is to be welcomed, it should not serve to reinforce distorted ideas of maternal agency. This issue is the focus of the next section.

13.4. AGENCY IN EVOLUTIONARY REPRODUCTIVE IMMUNOLOGY

Medawar's evolutionary thesis that selection pressures are driving fetuses toward independence shares with contemporary evolutionary explanations of pregnancy— both within reproductive immunology and without—its focus on maternal-fetal conflict. Some evolutionary accounts of pregnancy outside of immunology focus almost exclusively on maternal-fetal conflict (Haig, 1993). Robert Trivers (1974) model of parent-offspring conflict applied to pregnancy also centralizes conflict. In such views, the fetus will do all it can to avoid spontaneous abortion and to gain nutrition from the mother: the fetus is like a parasite, invading her tissues and altering her physiology. The mother, on the other hand, will attempt to restrict fetal nutrition and cut her losses by eliminating embryos not healthy enough to survive. Consider the claim that infectious abortion makes sense 'in evolutionary terms, since the mother is already of reproducing age and it is better for her to survive and reproduce again at the expense of the fetus' (Entrican, 2002, p. 90). That maternal survival and reproduction is thought to occur at the *expense* of the fetus highlights the maternal-fetal conflict at the heart of pregnancy from both immune and evolutionary perspectives. In what sense is the mother reproducing again at the *expense* of the earlier fetus? Maternal activity and subsequent success is seen to come at fetal expense, but if the mother dies, so will the fetus. And, while maternal immunological defenses against pathogens may harm the fetus, the fetus would likely suffer as a result of the infection anyway. Since the fetus 'loses' either way, it is hardly at its expense that the mother survives. This example helps illustrate how neatly the evolutionary focus on maternal-fetal conflict fits the foreign fetus model of immunology.

The fit between evolutionary models outside of immunology and those within, however, is not always so seamless. In reproductive immunology, recurrent pregnancy loss is often understood as a pathological process and not, for example, as a healthy response to immunologically incompatible sperm or an unhealthy fetus, as could be imagined in evolutionary accounts. As Jane Salmon illustrates the case in reproductive immunology:

The journey from conception to birth is fraught with danger. It has been estimated that 50–70% of all conceptions fail and that recurrent pregnancy loss affects 1–3% of couples ...When well-established genetic, anatomic, endocrine, and infectious causes of fetal damage are not demonstrable, as is the case in a majority of pregnancy complications, abnormal maternal immune responses are assumed to act as initiators of disease. (Salmon, 2004, p. 15)

We are dealing here with evidence of abnormality, not adaptive responses geared towards reproductive success.

But regardless of whether immunological and evolutionary accounts treat recurrent pregnancy loss as normal or pathological, they generally contain troublesome assumptions about maternal agency. As is now well known, gender-based assumptions occasionally stand in for evidence in evolutionary biology and available evidence is often understood in an androcentric manner. This is no less the case in evolutionary explanations of motherhood (Hrdy, 1999). Because negative assumptions about maternal agency influence evolutionary views of pregnancy, evolutionary approaches likely play a role in the marginalization of research concerning beneficial maternal immunity. My objective in this section is to show that because maternal-fetal conflict is the central focus of both evolutionary and reproductive immunological accounts of pregnancy, positive forms of maternal agency are inappropriately ignored.

To defend these claims, I will consider three examples of evolutionary explanation in the area of reproductive immunology. The first example concerns how pregnancy immunology is perceived given the longstanding division of immune functions into 'innate' and 'adaptive' categories. In the second example, I examine the proposal that pre-eclampsia/eclampsia is caused by an immunological 'unfamiliarity' with impregnating sperm—a situation thought by one research group to be particularly relevant to women with successive pregnancies fathered by different men. The third example concerns the lack of scientific interest in the immunological hypothesis that menstruation is an adaptation to infection. This issue relates to pregnancy immunology insofar as it shows how research concerning pathogen defense has been restricted to the realm of sex and ignored in menstruation and (to a lesser extent) pregnancy, in effect setting up a strange dichotomy between sex and reproduction in immunological research.

The first example deals with the claim that new that light may be shed on pregnancy immunology by examining innate immunity as opposed to the more traditional focus on adaptive (also known as acquired or specific) immunity.[7] Innate immunity is found in both invertebrates and vertebrates and is thus phylogentically ancient. It acts as a basic line of resistance against pathogens and includes a wide variety of general defenses. Various types of immune cells, including monocytes, macrophages, granulocytes, natural killer cells, and possibly $\gamma\delta$ T cells participate in these general defenses as do various antimicrobial substances. These innate means of resistance are not considered adaptive in the sense that they do not target specific pathogens with specialized knowledge acquired through adaptation of the immune response. Innate immunity is a general, non-specific, all-purpose defense system.

Adaptive immunity, on the other hand, is more specific in its defense and is responsible for phenomena such as the immunological memory of pathogens previously encountered. The immune system develops specialized responses to pathogens by learning through exposure what to attack and what to ignore. It is for this reason that adaptive immunity is related to self-nonself discrimination, and in turn, is held responsible for the specific identification of the fetus as foreign (Moffett-King, 2002).

Adaptive immunity has only evolved in vertebrates and is thus more phylogentically recent than innate immunity. It is perhaps for this reason that immunologists—including reproductive immunologists—emphasize adaptive immunity at the expense of innate immunity. Jan Klein (1999) says that until recently,

very few [immunologists] have been troubled by the neglect of the innate (nonanticipatory, nonadaptive, natural) form of immune response. I don't know how many immunologists there are in the world but I think the number is substantial, perhaps in excess of 10,000. Yet only a few of these seem to have realized that there is still a huge piece missing from the puzzle of immune resistance. (Klein, 1999, p. 488)

There are a number of reasons to think that human pregnancy might involve innate rather than adaptive immunity. For example:

The absence of B cells and paucity of T cells in the uterine mucosa suggests that the adaptive immune system is not as important as the innate system. Indeed, NK cells and macrophages are the main types of maternal leukocyte present. (Moffett and Loke, 2004, p. 4)

And, far from being immunosuppressed, indolent or inert, pregnant women have highly activated innate immune systems (Sacks et al., 1999; Entrican, 2002, p. 84). As Gary Entrican says, there is 'evidence that there may be alterations in maternal immune regulation during pregnancy, but this does not equate with immunosuppression' (Entrican, 2002, p. 84).[8]

Some think that the focus on innate immunity helps to resolve the immunological paradox of pregnancy. Since it is the adaptive immune system that aggressively and specifically targets nonself entities like transplanted tissue, if it is downregulated or turned off, even locally, aggressive activity need not be a problem. Recognition of the nonself fetus in human pregnancy, then, may actually have more in common with the sort of nonself recognition that occurs in invertebrates (Loke and King, 1997). Here, an explanation for safe maternal recognition of the fetus is achieved by treating pregnancy as a more phylogenetically ancient immune phenomenon.

From the perspective of maternal agency, however, this sort of explanation raises a concern. In immunology, some consider innate immunity more 'primitive' than adaptive immunity, or as one immunologist drolly laments, 'unsophisticated, unintelligent, indiscreet, and obsolescent' (Fearon, 1997, p. 323).[9] Thus, if pregnancy is an innate immune phenomenon, it would be considered more immunologically primitive. This assessment problematically intersects with Western cultural ideas that reduce aspects of pregnancy and childbirth to an instinctual or primitive level. Martin shows, for example, that some view birthing women as 'moving back in time and down the evolutionary tree to a simpler, animal-like, unselfconscious state' (Martin, 2001, p. 164). But as Martin argues, there are ways of seeing women as active participants in the birth process without assigning women primitive status—and one way is to recognize that the cultural and cognitive is part of the birth experience. Analogously, in the case of immunology, surely the fact that human beings are vertebrates—and vertebrates with a very unique pregnancy—makes a rather significant difference to how primitiveness is assigned. The integration between innate and adaptive systems must be taken into consideration.

It is also unclear whether the innate-adaptive distinction actually maps onto pregnancy. The evidence that it does not speaks against classifying pregnancy as immunologically primitive and in favour of classifying it as a unique immunological phenomenon. Consider the hypothesis that it may be the placenta-fetus that stimulates innate immunity. Under usual circumstances, infection activates the innate immune system. But, according to this hypothesis, the placenta-fetus provides the activation signal, not an infectious agent (Sacks et al., 1999). In fact, while

[i]mmunologists assert that the immune system has evolved to eliminate infection ...viviparous reproduction might have also influenced the evolution of the immune system. Lipid components on fetal-trophoblast membranes, perhaps in conjunction with other soluble circulating factors ...can generate

specific pregnancy signals through interaction with the innate system. Thus, the innate system might be able to distinguish the pregnant states from the non-pregnant state, producing a unique signal. (Sacks et al., 1999, pp. 116–117)

This unique signal may well bring into being a type of immune response specific to pregnancy—a type of response not equivalent to either innate or adaptive immunity in non-pregnant human beings. This seems likely, for there are some components of the innate immune system that are not activated during pregnancy, but are in cases of infection (Sacks et al., 1999, p. 116). Pregnancy, then, may have actively shaped immunological evolution rather than representing some kind of 'throw back.' But however innate immunity contributes to pregnancy, one thing is clear: the application of the innate-adaptive distinction to pregnancy does not move beyond the idea that pregnancy is an immune 'paradox.' It tries to avoid the paradox by limiting maternal immunological agency to the innate system, where it can do less harm. But there paradox is still there to be avoided.

The second example of problematic evolutionary explanation in reproductive immunology concerns women's sexual behaviour, specifically the sociobiological idea that women are more inclined to monogamy than men. Robillard et al. (2002) propose that pre-eclampsia/eclampsia has an immunological cause rooted in human female promiscuity or serial monogamy. Their reasoning is based on the 'fresh mating hypothesis,' which suggests that the risk of pre-eclampsia is highest in first pregnancies but declines in subsequent pregnancies, provided that the father remains the same. In other words, the more immunologically familiar a woman is with a man's foreign sperm, the less likely pre-eclampsia is to occur (though data from Basso et al., (2003) contradict this view). Robillard et al. (2002) propose that the need for immunological familiarity with sperm explains why systematic polyandry is rare and why there are cultural rules against it: pre-eclampsia strongly selects against polyandry.

Presumably, however, the benefits of consistent exposure to the same semen could just as easily develop in response to regular sexual activity with two men as one. A woman's immune system could become 'familiar'—if indeed this happens— with sperm from more than one male. In fact, it could be argued that systematic polyandry might reduce rates of pre-eclampsia: in a formal polyandrous relationship (as opposed to serial monogamy, for example), women would have regular exposure to the semen of more than one male, and hence, would have a reduced chance of contracting eclampsia in pregnancies resulting from any of their partners. In any case, cultural prohibitions of systematic polyandry would likely be insufficient to reduce rates of pre-eclampsia. The absence of formal polyandrous social arrange-ments does not rule out informal arrangements that make sexual variety possible for women. Any successful evolutionary immunological explanation should thus take into consideration women's *actual* sexual behaviour. Generally speaking, sexual variety was surely part of women's evolutionary past, passions being as they are, and it is certainly part of women's present.

The main difficulty with the evolutionary immunologically explanation of pre-eclampsia, however, may not involve assumptions about social arrangements, but the fact that other causes have been ignored: pre-eclampsia may be 'the extreme end of a continuum common to all pregnancies, with multiple contributing factors' (Redman and Sargent, 2003, p. S21). For example, one cause of pre-eclampsia may be sub-clinical infection (Viniker, 1999). Partner novelty may thus be relevant to pre-eclampsia not because of immunological unfamiliarity, but because a woman is exposed to new microbes or markedly increases her sexual activity. Both situations could result in an imbalance of the vaginal flora capable of influencing the progress of pregnancy.

The case of the fresh mating hypothesis shows that emphasizing maternal-fetal conflicts at the expense of infectious realities can lead to significant explanatory and empirical problems. The failure to consider adequately pathogen defense in pregnancy is also traceable to assumptions about the sexual agency of women. In this case, women are not seen as particularly active sexual agents (at least where variety is concerned), and pregnant women perhaps even less so. This creates problems for any model of pre-eclampsia that assumes women's monogamy: it simply is not safe to assume a stable microbial environment in evolutionary analyses of pre-eclampsia.

The third example of problematic evolutionary reasoning in reproductive immunology concerns the cool reception of Margie Profet's (1993) evolutionary explanation of menstruation as pathogen defense.[10] One explanation for the failure of relevant scientists to consider Profet's hypothesis seriously is Sharon Clough's view that the term 'female defense system' lacks salience in evolutionary biology— that is, it is not entrenched in linguistic usage. Clough contends this lack of salience explains why Profet's hypothesis has received little scientific attention, despite the fact that it plausibly brings menstruation and immune activities together in a coherent defense system (Clough, 2002, p. 721). Because the idea of an evolved female defense system is not available for thinking about menstruation, menstruation will not register as a possible adaptation to sperm-borne pathogens. This is compounded by the fact that menstruation is still generally considered as a non-functional byproduct of the fact that conception did not take place. Menstruation is thus quite easy to dismiss in adaptive explanations. Elsimar Coutinho and Sheldon Segal's (1999) theory that menstruation is obsolete is perhaps the best example of this. Martin argues that their theory contributed to the marginalization of Profet's theory—a theory that had the merit of treating 'menstruation as part of a woman's flexible immune system' (Martin, 2001, p. xvi).

In immunology, the situation is not much better. Generally, the menstrual cycle is studied with reference to allergy, anaphylaxis, and asthma in order to assess the influence of hormones on the severity of each condition. And, menstruation itself (as opposed to the cycle) is considered relevant in the immunological study of endometriosis. But little work has been done on the possible defensive functions of menstruation and there is reason to think it should be. Menstruation could be 'a major new site for immune function' (Martin, 2001, p. xvii). Immunologist David Clark suggests there may be unique immune cell types relevant to menstruation and considers this possibility in connection with defense against uterine infection (Clark, 1993; Clark and Daya, 1990). The types of immune cells present in the reproductive tract are known to change with the menstrual cycle and with pregnancy (Hunt and Soares, 1996). And, immunologist Ashley Moffett-King argues that uterine natural killer cells 'might have a dual role to play in reproduction, by monitoring mucosal integrity throughout the menstrual cycle and controlling trophoblast invasion during pregnancy' (Moffett-King, 2002, p. 657).[11] Moreover, the physical endometrial shedding involved in menstruation could also eliminate new infections by pathogenic bacteria, as is the case with shedding in other mucosal layers in the body.[12] Thus, there is increasing evidence of functional specialization in the immune cells of the female reproductive tract that reflects the requirements of both pathogen defense and pregnancy development.

In light of this, it is interesting that menstruation remains of marginal interest in reproductive immunology. By definition, we would expect reproductive immunology to concern female defense systems and for the concept to be more salient, to use Clough's term, in this area of biology than it is. However, the problem in immunology (and perhaps immunology differs from evolutionary biology in this respect), may not concern the salience of the 'female defense system' so much as the failure of this notion to apply conceptually to sex and pregnancy *together*. There is a very peculiar conceptual disconnection between sex and reproduction in immunological accounts of women's biology. The notion of female defenses against sperm-borne pathogens are typically associated with sex, not pregnancy. Mucosal immunology addresses sexually transmitted infections. Reproductive immunology, on the other hand, addresses the maternal-fetal relationship. From a theoretical standpoint, the infectious realities of sex, pregnancy and childbirth almost fade from view when a woman becomes pregnant. Pathogen defense is relevant to sexual activity. Maternal-fetal conflict is relevant to reproduction. Because menstruation is classified as reproductive, it is unlikely that it would be conceptualized as pathogen defense. However, menstruation is not thought relevant to maternal-fetal conflict management; it is merely a sign of pregnancy failure. Menstruation, therefore, has no clear disciplinary home.

That menstruation may be protective against infection suggests that the growth of the endometrium in women is not simply about providing nourishment for the blastocyst-fetus. I wonder, however, if this raises the possibility of a further function for menstruation. It seems possible that menstruation might act to shed immune cells that are functionally specialized to facilitate implantation— perhaps the cells functionally specialized for implantation are not as efficient at clearing infection as the usual immunological residents of the uterus. Menstruation may then reflect the convergence of several needs and have more than one function: pathogen defense, immunological preparedness for pregnancy, and the anatomical demands in human pregnancy for a sophisticated maternal-fetal interface. Recalling Martin's vision, functionally specialized immune populations that change with the menstrual cycle and pregnancy would indeed constitute a flexible immune system. While my hypothesis concerning the functionality of menstruation serves only as an example, it illustrates the sorts of hypotheses possible in an environment where beneficial reproductive immunological agency makes sense.

These examples of problematic evolutionary reasoning in reproductive immunology show that pathogen defense has been marginalized because of an overemphasis on maternal-fetal conflict and unary immunological selves. Fortunately, some immunologists are beginning to comment on the exclusion of pathogen defense in reproductive immunology. David Clark says that reproductive immunology 'has tended to focus on the immunological response to gametes and embryos without taking the broader view of immune responses as an adaptive strategy to deal with threats in the environment' (Clark, 1999b, p. 529). In a similar vein, Gary Entrican asks: 'What do these theories [the self-nonself and danger theories] predict would be the outcome of placental infection for the survival of mother or fetus, or both?' He then discusses evidence for immune mechanisms in pregnancy in the 'context of infectious abortion ...' (Entrican, 2002, p. 80). Other immunologists echo these views, though from the perspective of the foreign fetus model. Charles Wira and John Fahey remark that

the immune system must balance the presence of a resident population of bacteria in the vagina with periodic exposure in the uterus and Fallopian tubes of antigens (bacteria and sperm), as well as the long-term exposure of an allogenic fetus. Failure of the immune system either to rid the reproductive tract of potential antigens or to resist attacking allogenic sperm and fetus significantly compromises procreation, as well as the health of the mother. (Wira and Fahey, 2004, p. 13)

Similarly, Gavin Sacks, Ian Sargent and Christopher Redman ask, 'How can the fetus suppress maternal rejection responses but maintain, or even increase, her resistance to infection?' (Sacks et al., 1999, p. 115)[13] These remarks suggest that the general biological context of pregnancy is increasingly on the minds of some reproductive immunologists. What is needed now is a more developed theoretical and practical consideration of this context; the key is to find out how immunological contributions to pathogen defense and pregnancy work together. To do this, it would help to supplant the inert or pathologically active maternal immune system with a more robust and balanced view of maternal agency.

This more balanced view may be on the horizon. Moffett and Loke contend that the maternal immune system actively constructs the placental environment in a way that stimulates tissue growth and healing. Moreover, they argue that

[I]t is particularly important to move away from the view that the trophoblast is like a conventional allograft that must resist rejection. Instead, we should consider that the maternal immune response may be providing a nurturing balanced environment which curbs excessive or unsocial behaviour by both placenta and mother leading to a state of peaceful co-existence between two allogenic tissues (Moffett and Loke, 2004, p. 7).

What is interesting here, however, is the discordance between their call for viewing the maternal contribution as nurturing and balanced and their concerns about curbing 'excessive and unsocial behaviour.' If one truly eliminates the idea that pregnancy is an immunological paradox, does one really need to worry about restraining some unsocial immunological behaviour? Such pathology may signal a failure in the usual mechanisms of pregnancy immunology, rather than the release of some hostile force essential to pregnant women. Why should contact between two allogenic tissues necessarily cause problems? Pregnancy is not organ transplantation. It is actually quite disanalogous to organ transplantation when you think about it. So, despite the exciting work being done on this new more balanced view of the maternal-fetal immunological relationship, the tensions attending the foreign fetus model persist.

13.5. CONCLUSION: THE IMMUNOLOGICAL PARADOX OF PREGNANCY REVISITED

Conflict has been central to the understanding of maternal-fetal relations in immunological and evolutionary accounts of pregnancy. I contend that the centrality of conflict is largely due to a questionable ontology that creates a neat conceptual division between mother and fetus and then reifies it. The ontology that posits mothers and their fetuses as two distinct beings in conflict makes it easy to conceptualize the fetus as invasive and the mother as passively invaded—or to view the mother as a pathological immune agent destructively oriented towards the fetus. The static and territorial ontology behind the foreign fetus model is thus an empirically troubled space of foreign invaders, habitual aborters, cancer and parasitism. And, in emphasizing maternal antagonism over more positive forms of maternal agency, it has been easier to ignore the wider biological circumstances in which women's sexual activity and reproduction takes place.

My overall objective is to encourage immunology to consider explicitly models that incorporate better assumptions about maternal agency, ones that more clearly reflect the biological and social realities in which women find themselves. A conceptual change is needed to shift the centre of theoretical and experimental work in reproductive immunology away from passive and pathological understandings of maternal agency. An immunological model of the maternal-fetal relationship resting on a more socialized concept of agency might provide enough space in which to consider maternal immunity as active and beneficial *simultaneously*. The foreign fetus model cannot make sense of the findings that cooperative immunological interrelations are important for a healthy pregnancy (Ober and Van der Ven, 1996; Chaouat, 1993). Models explicitly including beneficial relationships can. It is important, then, to expand understandings of maternal immunological agency to include potentially protective, transformational, and cooperative activities, and to investigate these potential activities empirically. It may turn out that in taking a more expansive view of maternal immunological agency, pregnancy poses no paradox.

Trent University, Peterborough, Ontario, Canada

NOTES

[1]   Sargent, I., quoted in H. Pearson 'Immunity's Pregnant Pause,' 21 November 2002, *Nature Science Update*, http://www.nature.com/nsu/

[2]   There are, of course, many reasons to study sex differences in autoimmune disease. Approximately 80% of the 8.5 million people with autoimmune diseases in the U.S. are women and as of yet, there is no clear answer as to why this is (Whitacre et al., 1999). Despite the social, personal and economic impact of these diseases, specially designated funds from the NIH for studying gender differences in autoimmune disease have only recently become available (Whitacre, 2001).

[3]   See, for example, Moffett and Loke, 2004; Van Nieuwenhoven, Heineman and Faas, 2003.

[4]   An antigen is any molecule that can be recognized by adaptive components of the immune system.

[5]   I consider this issue in section 13.3 of the paper.

[6]   My thanks to Emily Martin for emphasizing the importance of sociality to my discussion of agency.

[7]   I do not address here the increasingly controversial issue of whether a genuine distinction between innate and adaptive immunity actually exists.

[8]   It is thought that in pregnancy there is a shift in levels of cytokines produced by T-helper 1 lymphocytes and T-helper 2 lymphocytes. The view that Th1 vs. Th2 bias occurs—though this is still controversial—could also be understood as a change in regulation rather than as immunosuppression. See Gleicher, 2002; Croy, 2000; Wegmann et al., 1993.

[9]   The $\gamma\delta$ T cell, for example, lacks the specificity of T lymphocytes of the adaptive immune system and this has led to the view that it is more 'primitive' than other T cells (Entrican, 2002, p. 82).

[10]   Beverly Strassman (1996) offers a critique of Profet's explanation. There are problems with both Profet and Strassman's accounts of the functionality of menstruation; however, space does not permit discussion of these here.

[11]   Note the use of the invasion concept.

[12]   My thanks to Richard Cone, Department of Biophysics, Johns Hopkins University, for pointing this out to me.

[13]   Here, the fetus is controlling the maternal immune system. The maternal is passively inactive.

REFERENCES

Adachi H, Takakuwa K, Mitsui T, Ishii K, Tamura M, Tanaka K (2003) Results of immunotherapy for patients with unexplained secondary recurrent abortions. Clin Immunol 106:175–180

Basso O, Weinberg CR, Baird DD, Wilcox AJ, Olsen J (2003) Subfecundity as a correlate of preeclampsia: a study within the Danish national birth cohort. Am J Epidemiol 157:195–202

Ben-Chetrit A, Ben-Chetrit E (1994) Systemic lupus erythematosus induced by ovulation induction treatment. Arthritis Rheum 37:1614–1617

Bianchi DW (2000) Fetomaternal cell trafficking: a new cause of disease? Am J Med Genet 91:22–28

Birke L (1999) Feminism and the biological body. Edinburgh University Press, Edinburgh

Bonney EA, Matzinger P (1997) The maternal immune system's interaction with circulating fetal cells. J Immunol 158:40–47

Bubanovic I, Najman S (2004) Failure of anti-tumor immunity in mammals – evolution of the hypothesis. Acta Biotheor 52(1):57–64

Bussen S, Steck T (1997) Thyroid antibodies and their relation to antithrombin antibodies, anticardiolipin antibodies and lupus anticoagulant in women with recurrent spontaneous abortions (antithyroid, anticardiolipin and antithrombin autoantibodies and Lupus anticoagulant in habitual aborters). Eur J Obstet Gynecol Reprod Biol 74:139–143

Casoli P, Tuniati B, La Sala G (1997) Fatal exacerbation of systemic lupus erythematosus after induction of ovulation. J Rheumatol 24:1639–1640

Chaouat G (1993) The roots of the problem: 'The Fetal Allograft,' In: Chaouat G (ed) Immunology of pregnancy, CRC Press, Boca Raton, pp 1–17

Clark DA (1993) Uterine cells of immunological relevance in animal systems, In: Chaouat G (ed) Immunology of pregnancy, CRC Press, Boca Raton, pp 79–92

Clark DA (1999a) Signaling at the fetomaternal interface. Am J Reprod Immunol 41:169–173

Clark DA (1999b) Hard science versus phenomenology in reproductive immunology. Crit Rev Immunol 19:509–539

Clark DA, Arck PC, Chaouat G (1999) Why did your mother reject you? Immunogenetic determinants of the response to environmental selective pressure expressed at the uterine level. Am J Reprod Immunol 41:5–22

Clark DA, Daya S (1990) Macrophages and other migratory cells in endometrium: relevance to endometrial bleeding, In: D' Arcangues C, Fraser I, Newton J, Odlind V (eds) Contraception and mechanisms of endometrial bleeding, Cambridge University Press, Cambridge, pp 363–382

Clough S (2002) What is menstruation for? On the projectibility of functional predicates in menstruation research. Stud Hist and Philos Biol Biomed Sci 33:719–732

Coutinho E, Segal S (1999) Is menstruation obsolete? Oxford University Press, New York

Creus M, Balasch J, Fábregues F, Martorell J, Boada M, Penarrubia J, Barri PN, Vanrell JA (1998) Parental human leukocyte antigens and implantation failure after in-vitro fertilization. Hum Reprod 13:39–43

Croy BA (2000) Where now for the Th1/Th2 paradigm of the gestational uterus? J Reprod Immunol 51:1–2

Entrican G (2002) Immune regulation during pregnancy and host-pathogen interactions in infectious abortion. J Comp Pathol 126:79–94

Fearon D (1997) Seeking wisdom in innate immunity, Nature 388:323–324

Geenen V, Perrier de Hauterive S, Puit M, Hazout A, Goffin F, Frankenne F, Moutschen M, Foidart JM (2002) Autoimmunity and pregnancy: theory and practice. Acta Clin Belg 57:317–324

Gleicher N (2002) Some thoughts on the reproductive autoimmune failure syndrome (RAFS) and Th-1 versus Th-2 immune responses. Am J Reprod Immunol 48:252–254

Govallo VI (1993) The immunology of pregnancy and cancer. Nova Science Publishers, Commack

Haig D (1993) Genetic conflicts in human pregnancy. Q Rev Biol 68:495–532

Hall J (2003) So you think your mother is always looking over your shoulder? – She may be in your shoulder! J Pediatr 142:233–234

Haraway D (1989) The biopolitics of postmodern bodies: determinations of self in immune system discourse. Differences 1:3–43

Hunt JS (ed) (1996) HLA and the maternal-fetal relationship. R. G. Landes Company, Austin

Hunt J, Soares M (1996) Features of the maternal-fetal interface, In: Hunt JS (ed) HLA and the maternal-fetal relationship, R. G. Landes Company, Austin, pp 1–26

Hrdy SB (1999) Mother nature: a history of mothers, infants, and natural selection. Pantheon Books, New York

Kallen C, Arici A (2003) Immune testing in fertility practice: truth or deception? Current opinion in obstetrics and gynecology 15:225–231

Karpin I (1992) Legislating the female body: Reproductive technology and the reconstructed woman. Columbia Journal of Gender and Law 3:325–333

Katz I, Fisch B, Amit S, Ovadia J, Tadir Y (1992) Cutaneous graft-versus-host-like reaction after paternal lymphocyte immunization for prevention of recurrent abortion. Fertil Steril 57:927–929

Klein J (1999) Immunology at the millennium: Looking back. Curr Opin Immunol 11:487–489

Koch CA, Platt JL (2003) Natural mechanisms for evading graft rejection: The fetus as an allograft, Seminars in Immunopathology 25:95–117

Langman R (ed) (2000) Self-nonself discrimination revisited, Seminars in Immunology, 12

Langman R, Cohn M (2000) A minimal model for the Self-nonself discrimination: a return to the basics, Seminars in Immunology 12:189–195

Laskin CA, Chuma A, Angelov L, Neil G, Levy GA, Mason N, Soloninka C, Cole E (1994) Sera from habitual aborters induce monocyte procoagulant activity: a lymphocyte-dependent event. Clinical Immunology and Immunopathology 73:235–244

Loke YW, King A (1997) Immunology of human placental implantation: clinical implications of our current understanding. Mol Med Today 3:153–159

MacKenzie C, Stoljar N (eds) (2000) Relational autonomy: Feminist perspectives on autonomy, agency and the social self. Oxford University Press, Oxford

Martin E (1994a) Flexible bodies: Tracking immunity in American culture from the days of polio to the age of AIDS. Beacon Press, Boston

Martin E (1994b) The ethnography of natural selection in the 1990s Cultural Anthropology 9:383–397

Martin E (2001) The woman in the body: a cultural analysis of reproduction. Beacon Press, Boston

Matzinger P (1994) Tolerance, danger, and the extended family. Annu Rev Immunol 12:991–1045

McLeod C (2002) Self-trust and reproductive autonomy. MIT Press, Cambridge, Mass.

Medawar PB (1953) Some immunological and endocrinological problems raised by the evolution of viviparity in vertebrates. Symposia-Society for Experimental Biology 44:320338

Moffett-King A (2002) Natural killer cells and pregnancy. Nature Reviews Immunology 2:656–663

Moffett A, Loke YW (2004) The immunological paradox of pregnancy: A Reappraisal, Placenta 25:1–8

Nelson JL (2001) Microchimerism: expanding new horizon in human health or incidental remnant of pregnancy? Lancet 358:2011–2012

Nelson JL (2002) Microchimerism: incidental byproduct of pregnancy or active participant in human health? Trends Mol Med 8:109–113

Ober C, Van der Ven K (1996) HLA and fertility, In: Hunt JS (ed) HLA and the maternal-fetal relationship R. G. Landes Company, Austin, pp 133–156

Profet M (1993) Menstruation as defense against pathogens transported by sperm, Q Rev Biol 63:335–386

Redman C, Sargent I (2003) Pre-eclampsia, the placenta and the maternal systemic inflammatory response—a review. Placenta, 24(Supplement 1):S21–S27

Roberts RM, Xie S, Mathialagan N (1996) Maternal recognition of pregnancy. Biol Reprod 54:294–302

Robillard P-Y, Dekker GA, Hulsey TC (2002) Evolutionary adaptations to pre-eclampsia/eclampsia in humans: low fecundability rate, loss of oestrus, prohibitions of incest and systematic polyandry. Am J Reprod Immunol 47:104–111

Sacks G, Sargent I, Redman C (1999) An innate view of human pregnancy. Immunol Today 20:114–118

Salmon J (2004) A non-inflammatory pathway for pregnancy loss: innate immune activation? J Clin Invest 114:15–17

Sherwin S (1998) A relational approach to autonomy and health care, In: Sherwin S (Coordinator), The politics of women's health: exploring agency and autonomy, Temple University Press, Philadelphia, pp 19–47

Strassman B (1996) The evolution of endometrial cycles and menstruation. Q Rev Biol 71:181–220

Sumner LW (1981) Abortion and moral theory. Princeton University Press, Princeton

Tanaka T, Umesaki N, Nishio J, Maeda K, Kawamura T, Araki N, Ogita S (2000) Neonatal thrombocytopenia induced by maternal anti-HLA antibodies: a potential side effect of allogenic leukocyte immunization for unexplained recurrent aborters. J Reprod Immunol 46:51–57

Tauber A, Podolsky S (1997) The generation of diversity. Harvard University Press, Cambridge, Mass.

Trivers R (1974) Parent-offspring conflict. American Zoologist 14:249–264

Van Nieuwenhoven V, Heineman MJ, Faas MM (2003) The immunology of successful pregnancy. Hum Reprod Update 9:347–357

Viniker DA (1999) Hypothesis on the role of sub-clinical bacteria of the endometrium (bacteria endometrialis) in gynaecological and obstetric enigmas. Hum Reprod Update 5:373–385

Weasel L (2001) Dismantling the self/other dichotomy in science: towards a feminist model of the immune system. Hypatia 16:27–44

Wegmann TG, Lin H, Guilbert L, Mossman TR (1993) Bi-directional cytokine interactions in the maternal-fetal relationship: is successful pregnancy a Th2 phenomenon? Immunol Today 14:353–356

Whitacre C (2001) Sex differences in autoimmune disease. Nature Immunology 2:777–780

Whitacre C, Reingold S, O'Looney P, The task force on gender, multiple sclerosis and autoimmunity (1999) A gender gap in autoimmunity. Science 283:1277–1278

Wira C, Fahey J (2004) The innate immune system: gatekeeper to the female reproductive tract. Immunology 111:13–15

JONATHAN KAPLAN

14. VIOLENCE AND PUBLIC HEALTH: EXPLORING THE RELATIONSHIP BETWEEN BIOLOGICAL PERSPECTIVES ON VIOLENT BEHAVIOR AND PUBLIC HEALTH APPROACHES TO VIOLENCE PREVENTION

14.1. INTRODUCTION: BIOLOGY AND VIOLENT BEHAVIOR

At least two broad approaches to research aimed at understanding the relationship between human biology and violent behavior still have some currency. The 'classic' view starts from the observation that cross-culturally a relatively small percentage of young men commit the majority of violent crimes, and has as its goal finding a way to identify these men before their violent criminal careers begin. A closely related but relatively more recent approach argues that individuals become violent not because of particular biological dispositions or environmental stressors, but rather due to the combination of biological and environmental influences; it takes as its goal finding a way to identify those individuals 'at risk' for violent behavior so that their (likely) future behavior can be changed. In the second broad approach, researchers have argued that violent behavior is a non-pathological response to particular environmental triggers, and takes as its goal finding a way to modify people's developmental environments so that they are not exposed to those triggers. While none of these programs (or related variants of them) has yet produced any unequivocal results, all promise that by helping us to understand the biological determinants of violent behavior, they will help us to reduce the prevalence and impact of violence in society. As, at least among the relatively young, violence is one of the leading causes of mortality and morbidity world-wide, the temptation of this promise is obvious.

But the promises of these programs are also deeply misleading. As Ian Hacking points out, these approaches to violence are part of a more general approach to the medicalization of criminal and violent behavior, and that approach has a long history of failure (Hacking, 2001; Kevles, 1985). More to the point, even if the current research programs succeed on their terms, they cannot in fact yield results that will be useful for the amelioration of violence or the causes of violence. This is not to say that the results of such research programs, if successful, would necessarily be uninteresting; rather, any interest in the results would have to come *not* from any policy implications of the results but rather from the role they would play in some more general explanatory project.

In what follows I will briefly outline each general approach (including some of the obvious variants of them), what the proponents of each approach take to

199

H. Kincaid and J. McKitrick (eds.), Establishing Medical Reality, 199–214.
© 2007 *Springer*.

be compelling evidence that they are on the right track, and why none of these approaches are likely to yield results that will have any public policy implications that rely on the details of the research results. But first, it is worth pausing to consider what kinds of results public policy changes ought to aim at – what, in other words, we might hope to accomplish by studying violence and the causes of violence.

14.2. VIOLENCE AND PUBLIC HEALTH

The reasons for taking steps to reduce the prevalence and ameliorate the effects of violence are obvious; interpersonal violence is one of the leading causes of death and serious injury in children and young adults in both developed and developing nations, and there are very high social and economic costs associated with violence (Krug, Sharma and Lozano, 2000, pp. 523–526). The so-called 'public health' approach to violence suggests that since interpersonal violence is, in principle, preventable, the deaths, injuries, and costs associated with violence are likewise preventable (Mercy et al. 2003, pp. 256–261). Since the public health system is one of the social systems responsible for working to prevent preventable injuries and deaths, the argument continues, the public health system ought to be one of the social systems concerned with preventing violence (Freudenberg, 2000, pp. 473–503; Krug et al., 2000, pp. 523–526; Mercy et al., 2003).

This position is, admittedly, somewhat controversial; some have argued that the public health system is not the right social system (or one of the right social systems) through which to try to prevent violence (Elliott, 1998, pp. 1081–1098 and McDonald, 2000, pp. 1–6 for discussion). However, even if the goal of preventing the injuries and deaths associated with violence by reducing the prevalence (and ameliorating the effects) of violence is removed from the purview of public health professionals, many of the same kinds of programs, and arguments for such programs, are made by those in other arenas; hence, the importance of the debate over which groups should be responsible for the creation and/or advocacy of such programs is at least reduced.

In any event, public health approaches to reducing the prevalence and ameliorating the effects of violence in societies tend to focus on a number of social factors associated with violence rates. These include extensive economic (and more broadly social) inequality (Kennedy et al., 1996, pp. 1004–1007; 1998, pp. 917–921; Rodgers, 2002, pp. 533–538), lack of social and economic opportunities (Freudenberg, 2000; McCulloch, 2001, pp. 208–209), the availability of handguns (Lester, 1991, pp. 186–188; Rawson, 2002, pp. 379–383), and the like. Of course, the public health system is not the primary social/political system responsible for these factors; changing, for example, the level of social/economic inequality within a society demands structural changes outside the usual purview of the health-care community. But even here, proponents of a public-health approach to violence suggest that by advocating for such changes, and by keeping the costs of resisting such changes foregrounded, the public health sector can have a real impact on the likelihood of such changes being implemented (McDonald 2000; Guerrero 2002, p. 767).

Much more controversially, some authors have suggested that recognizing the biological bases for violent behaviors (including genetic associations) might play an important part of a public health approach to preventing violence and ameliorating the effects of violence (DiLalla and Gottesman 1991, pp. 125–129; Rsensberger 1992; Gibbs 1995, pp. 101–107; Caspi et al., 2002, pp. 851–854; Lowenstein 2002, pp. 767–772; Martens 2002, pp. 170–182). Some of these authors suggest that studying the biological bases of violent behavior might permit the full or partial medicalization of violence; the idea is that once we find the biological pathways that result in violent behavior, we might be able to intervene at the level of biochemistry (Brunner et al., 1993, pp. 578–580; Caspi et al., 2002, Martens, 2002). This might take the form of more or less traditional medical treatment (e.g., drugs, surgeries, etc.), or it might take the form of specific social interventions. But (medical) intervention will be, these researchers imply, required; some researchers have suggested that society 'must identify high-risk persons at an early age and place them in treatment programs *before* they have committed' any crimes (Jeffery, quoted in Gibbs, 1995). Or, as another researcher has suggested, once you can show that a person is likely to commit crimes for biological reasons and hence is ill in the traditionally medical sense, they should be treated as medically ill, and treatment, if available, 'should be mandatory,' or, if no treatment is currently available, then the biologically (potential) recidivist criminal 'should be held indefinitely' (Fishbein, quoted by Gibbs 1995).

The stated aim of both the more traditional public health approaches to violence and those approaches based on biological correlates is, again, to reduce the prevalence of violence in societies and to ameliorate the effects of the violence that does occur. These goals are considered feasible in part because of the wide range of rates of violence found in different societies and the fact that violence rates in particular locations change over time (Day, 1984, pp. 917–927; Krug et al., 1998, pp. 214–221; Barclay et al., 2002). Since there are nations (and individual cites within nations) that have relatively low rates of violent crime and relatively low rates of injuries caused by violent behaviors, the reasoning goes, it should be possible to modify those social settings with relatively high rates of violent crime and injuries caused by violence to reduce their rates of violent crime and injuries caused by violence to rates more like those found in social settings with relatively low-rates. Since it is widely recognized that violent crime (and violence more generally) has very high costs associated with it – for example, the costs associated with medical treatments, the lost income from injuries and deaths, the costs to the criminal justice system (the cost of policing, investigations, trials, incarceration, etc.), and the costs associated with the physical pain and emotional suffering – it could easily be worth using considerable social resources to try to reduce the violent crime rates, even if those projects attempted resulted in relatively modest gains.

14.3. BIOLOGICAL APPROACHES TO UNDERSTANDING THE CAUSES
OF VIOLENT CRIME AND VIOLENCE: PROGRAMS, PROMISES,
AND PROBLEMS

Attempts to understand and control violent and/or criminal behavior through
biological approaches have a long history (Gould, 1981; Kevles, 1985; Hacking,
2001); however, despite some of the bold claims by early proponents of the biolog-
icalization of violent criminals (such as those made in the late nineteenth and
early twentieth centuries by Lombroso and his student Ferri) these approaches
never met with any real empirical success (see Gould, 1981 22ff; Lewontin et al.,
1984 55ff). This has not, though, much dampened the enthusiasm with which the
potential to find biological markers associated with violent and criminal behavior
is met (Hacking 2001). For example, Moffitt and Mednick (1988) claim that since
'social and environmental variables cannot explain' those whose violent criminal
careers start early and continue throughout life, a biological approach is necessary;
if such an approach could recognize those '5% of males [who] commit over 50% of
criminal offenses' we could 'dramatically reduce our growing crime rate' by 'intervention directed at these relatively few individuals' (Moffitt and Mednick, 1988;
Gibbs, 1995). Breakefield suggests that early biochemical testing and intervention
might be an important part of a prevention program, because with many kind of
deficiencies, 'if you don't change the diet within the first year, it's too late,' and
while, for example, MAOA deficiency (discussed below) 'does not automatically
lead to violence, or indeed any other kind of behavior,' preventing it from doing
so might well require medical intervention (Breakefield, quoted in Mann 1994,
pp. 1686–1689). Or, as Hamer and Copeland put it, the 'real reason to study the role
of genes and biology in aggression is to understand what can be changed and what
cannot, what works and what doesn't' (1998, pp. 126–7); a biological approach to
violent crime is, these authors suggest, the best way of finding effective ways of
prevent its occurrence and/or ameliorate its effects. But how realistic is this view?

As noted above, there are currently two broad approaches to attempting to understand (and to control) violence (and especially criminal violence) through biological
models. The first approach relies on attempting to find biological markers associated
with increased risks of violent (and/or criminal) behaviors in individuals; it starts
from the observation that individuals within a population seem to differ with respect
to how likely they are to commit violent crimes (or act violently more generally),
and attempts to find biological features of the individuals associated with those
different likelihoods. Since it is well-established that there are *environmental* (social,
socioeconomic, etc.) correlates with violent crime, a contemporary variant of this
kind of research looks not for biological markers that explain the variance in violent
behavior *per se* but rather towards interactions between biological markers and
environmental influences that can explain the variance in behavior. The second
broad approach treats violence (and criminal violence) as an universal adaptive
response to particular environmental circumstances, and posits that particular
environmental conditions will trigger the development of violent personality types in
any normally developing human. The variation with respect to violence both within

and between populations, then, must be the result of differences in the frequency of these environmental triggers. Other researchers interpret the data as suggesting that there might be multiple normal 'types' of people, adapted to different local conditions, and hence that the same 'normal' development conditions will therefore lead some people to be more violent than others; this, the researchers claim, can better explain cross-cultural variations in the rates of violence and violent crimes.

While I believe that none of these approaches have generated evidence that is without serious empirical difficulties (Sober, 2001; Schaffner, 2001), in what follows I will focus instead on critiquing the claim that current and foreseeable biological approaches to understanding violence could be an integral part to a social strategy to reduce the prevalence and impact of violence and violent crime in society. Even if the finding reported by, and the arguments made by, proponents of these approaches are on the right track, I will suggest that they do not add anything to our current understanding of the best ways to reduce the prevalence and ameliorate the impact of violence and violent crime in our societies. That is not to say that these research programs won't provide interesting insights of other sorts – if indeed they are able to overcome the serious problems with research into the biological bases of complex human behaviors (Kaplan, 2000; 2002, and cites therein), research into the biological correlates and bases of violence might provide some fascinating insights into human violence. However, I suggest here that these insights are very unlikely to provide much or any guidance regarding the public policies most likely to be successful in reducing violent crime, interpersonal violence, and the effects violence.

14.3.1. 'Traditional' Behavior Genetics: A Focus on Individual Variation

One of the most oft-cited statistics among those hoping to find ways to identify those people likely to commit violent crimes before they have done so (or before, at least, they have committed many violent crimes) is that, cross-culturally, a small fraction of young men (between around 6% and 10%) commit the majority of violent crimes (between around 50% and 70%) (see Wilson 1994, pp. 25–35; Gibbs 1995, UK Home Office Report 2001 Annex B; see also Caspi et al., 2002). If some way could be found to identify those people likely to become the violent repeat offenders (or identify them very early in their criminal careers) and prevent them from doing so, the argument continues, the prevalence of violent crime could be reduced quite drastically. Since most young men do not become violent criminals, something must be different about those that do; given that in many cases the violent criminals seem to come from 'the same' social environment as those that do not become violent criminals, a search for biological markers is reasonable. This line of argument was made most explicitly in Mark and Ervin's 1970 book *Violence and the Brain.* Today most authors arguing for genetic or other biological links to individual variation in violent criminal behavior focus on estimates of the heritability of violent behaviors or searches for genetic markers associated with violent criminal behaviors; it is worth noting, however, that these kinds of research programs only makes sense if the kind of argument outlined above is accepted.

From a public health standpoint, the most serious problem with the above line of argument is that it fails to address the different rates of violence *between* societies. Even if it is true that cross-culturally a small fraction of young men commit the majority of the violent crimes, the *rate* of violent crime (as well as the relative frequency of the different types violent crimes) varies enormously between societies. While comparisons of violent crime rates between nations is difficult (see Barclay and Tavares, 2002), it seems certain that there is at the very least a ten-fold, and likely a twenty-fold or greater, spread of homicide rates between countries with relatively low rates (e.g., around 1–2 homicide deaths per 100,000 people in Japan, Norway, Finland, Canada, etc) and those with relatively high rates (e.g., around 20–50 homicide deaths per 100,000 in Mexico, Russia, and South Africa, and around 6–10 per 100,000 in the United States) (see e.g. Day, 1984; Barclay and Tavares, 2000; Krug et al., 2002). It is much harder to generate accurate statistics regarding assaults that produce non-lethal injuries, but it seems that there is similarly a very large spread between cultures with the most frequent assaults and those with the least frequent assaults. Reliable estimates of the rates of sexual assault are for a variety of reasons even more difficult to come by (see for example Dussich et al., 1996) but again, reasonable estimates of the rates in different countries imply a very wide spread (see Krug et al., 2002, 88ff., 147ff.).

So even if, as seems very unlikely, we did find a way to identify that small fraction of men who commit most of the violent crimes in any given society and to prevent them from doing so, the extent to which we had address the public health aspect of violent criminal behavior would be determined by the background rate of violent crime in the society in question. That is, if for example this project were undertaken in Russia or South Africa, even after all the potentially recidivist young men had been prevented from committing their crimes, the remaining homicide rate would *still* be over ten times that of Japan or Norway. Even if there are biological correlates to violent criminal behaviors that are shared by the small fraction of young men who cross-culturally commit the majority of violent crimes, the differences in the *rates* of violent crime must be independent of these correlates. So the supposed biological correlates of violent criminal behavior cannot be a useful part of the explanation of violent crimes rates, and hence discovering these correlates simply cannot be the most useful approach to reducing the rates of violent crime in those societies in which violent crime is most prevalent.

Here, one might be tempted to suggest that while not perhaps the most effective place to look for reductions in violent crime rates, finding and preventing that small fraction of men from committing those violent crimes would still be very useful – after all, such a program could, in principle, reduce the rate of violent crimes by around 50%. But of course, we cannot now, nor does it seem likely that we will soon be able to, identify those young men most likely to commit the majority of violent crimes before they have done so, or even particularly early in their criminal careers. Given that we currently have very little (if any) reliable information on just what causal biological pathways might be involved in 'making' this small fraction of young men into repeat violent criminals, it is of course very hard to guess about how effective potential interventions in these biological pathways might be, or, for that matter, even what form such interventions might take (see Schaffner, 2001 for a humbling account of how difficult it is to trace the biological pathways involved in even relatively simple behaviors in even very simple model organisms).

But we *can* identify some of the social factors responsible for the different rates of violent crime in different societies. 'Risk factors' for a society having a high violent crime rate include wide gaps in relative income/wealth, high levels of very low income more generally, poor or otherwise unreliable policing and judicial systems, poor educational and job opportunities for large groups of the population, and a variety of cultural factors relating to control and the perceived acceptability of violence (see e.g. Lester, 1986; 1991; Rockett, 1998; Krug et al., 1998, pp. 214–221; Barclay et al., 2001, etc.). And we *do* know, at least in very broad outlines, what kinds of interventions are likely to be effective. For example, programs that target 'at risk' youth and include parenting skills classes, multiple home visits, conflict-resolution training, and the like, have met with real success in reducing both the risk of children growing up to become violent adult criminals *and* in reducing domestic violence and child abuse within the targeted homes (see e.g. Kellermann et al., 1998, pp. 271–292; and cites therein, Mugford and Nelson 1996). Moves towards 'community policing' have shown promise in reducing the rates of violent crime in particular locations (see e.g. Cole 1999, 182ff. and cites therein), and increased educational and economic opportunities for women have shown real promise in reducing domestic violence via empowering women (see e.g. Sen 1999, especially chapter 8, and cites therein). More generally, increasing economic and educational opportunities have shown real promise in some studies (see e.g. Cole 1999, 182ff. and cites therein, Freudenberg, 2000 and cites therein).

It is easy to dismiss such proposals as politically unrealistic, too uncertain to succeed, too expensive, etc., and perhaps these objections are reasonable. However, such fatalism applies equally well to proposals built on the hope of finding biological correlates to violent criminal behavior. To suggest that because environmental interventions at the social level are too difficult, one should turn instead to biological approaches is likely simply a way of detracting attention from an unwillingness to even *try* to address the social concerns (see Taylor, 2001 for discussion).

A relatively concrete example may help make this clearer. Recently, attention has turned from 'simple' biological correlates with criminal behavior to trying to uncover the ways that biological and environmental variations interact to produce particular behavioral tendencies. In the early 1990's, a family in the Netherlands was found where many of the male members had a record of 'abnormal' behavior (including violent antisocial behaviors); biochemical testing revealed these men to be deficient in MAOA (Monoamine Oxidase A), and genetic testing revealed a nonsense point mutation in the MAOA gene; as the MAOA gene is X-linked, these men had no working MAOA gene, explaining their lack of MAOA (Brunner et al., 1993). But it rapidly became clear that complete MAOA deficiency was extremely rare, and studies attempting to link partial MAOA deficiency to aggressive antisocial behavior tended to be inconclusive (Caspi et al., 2002). However, when the effects of the early developmental environment were considered, a strong relationship was found between growing up in an abusive household and the likelihood of aggressive violent behavior later in life, an association made much stronger for (likely) low levels of MAOA activity. Children were categorized according to the likely level of physical abuse in their homes (the abuse was generally directed at the children and their mothers), from likely none, to probable and moderate abuse, to likely severe abuse; the likely MAOA level of the children was determined by the genotype of the MAOA promoter region they had, longer promoter regions being associated with greater MAOA activity (Caspi et al., 2002). While children who grew up in house-holds likely to have been non-abusive had the same (low) risk of becoming violent adults whatever version of the MAOA promoter region they had, those children who grew up in abusive households had a much greater risk of becoming violent adults if they had promoter regions associated with low-MAOA activity than if they had high-MAOA regions (see Figure 1).

Caspi et al., 2002 end their article with the claim that the high relative risk and predictive sensitivity of the association between low-MAOA, severely abused children with adult violent behavior 'indicate that these findings could inform the development of future pharmacological treatments' (2002, p. 853). Newspaper reports were quick to pick up on this thread, noting that the discovery of the

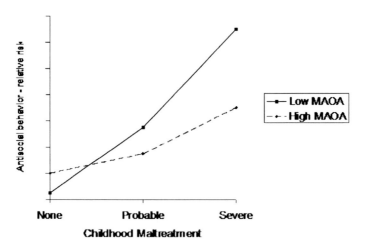

Figure 1. The relative risk of antisocial behavior for adults who grew up in households with no abuse, probable abuse, and severe abuse, for people with likely low levels of MAOA and those with likely high levels of MAOA. Note that children who grow up in households with physical abuse have a higher risk of becoming violent adults, whatever there MAOA levels, but that those with low MAOA levels have a higher risk given abuse than those with higher levels. Redrawn from (Caspi et al., 2002).

(On the concept of a 'generalized' norm of reaction see Sarkar and Fuller 2003.)

link between growing up in abusive households, low-MAOA activity, and violent behavior might lead to 'good behavior' pills (Hawkes, 2002). And some researchers have explicitly claimed that such links as Caspi et al. have found make the development of 'pharmacological' responses to violent criminal behavior both scientifically plausible and socially reasonable (Adam, 2004).

On the face of it, these responses to Caspi et al's findings would seem to be quite insane. Even if we could develop an easy 'treatment' for low-MAOA levels, MAOA is quite obviously the wrong place to intervene. Even if all the children with low-MAOA levels in Caspi et al's study were treated such that their MAOA levels were as high as the 'high' MAOA children, there would still be the problem that childhood abuse is associated with a greater risk of children growing up to become violent adults *whatever* their MAOA levels (see Figure 1). And, 'treating' a child with a low MAOA level who is growing up in an abusive household by *raising* that child's MAOA level while leaving the abuse in place would be horribly cruel even if it *were* completely effective vis-à-vis the child's risk of growing up to become a violent adult.

As it turns out, while we have no idea how to raise a developing child's MAOA level in any effective way, we *do* know quite a bit about how to prevent child and spouse abuse (see e.g. Mugford and Nelson, 1996 and cites therein, Rockett, 1998 and cites therein, Kellermann et al., 1998 and cites therein). Doing this would, if Caspi et al's finding hold up, reduce the risk of children growing up to become violent adults *whatever* their MAOA levels; indeed, if we succeeding in preventing child from growing up in abusive households, it *wouldn't matter* whether they had relatively high or relatively low MAOA levels (wouldn't matter with respect to their risk of growing up to be violent adults; it might matter for other conditions – see Winkelmann, 2003, pp. 81–82). And, while there is evidence that preventing children from growing up in abusive households would have other positive effects on their development (see Koenen et al., 2003, pp. 297–311), surely preventing children from growing up in abusive households is valuable in and of itself. *Whatever* the effect on the child's risk of growing up to be a violent adult, preventing that child from being severely abused and/or preventing that child from having to watch his or her mother being severely abused, is surely something worth working for (as of course is preventing spousal abuse more generally). And again, unlike modifying MAOA levels, there are strategies that have been shown to be effective in reducing domestic violence.

So, while there are many ways in which Caspi et al., results might prove to be less general than the above assumes, from a public health standpoint, the relationship between low-MAOA levels and violence is, at best, irrelevant, and at worst a dangerous distraction. This is not to say that it is *uninteresting* – should the results of Caspi et al. prove to be replicable and generalizable to other populations, we certainly will have learned something interesting about a particular kind of interaction between human developmental environments and genetic variation. And this finding might indeed lead into other interesting research agendas; we might, someday, look upon this kind of research as having opened one door into the complexities of human development and behavior. But none of this is relevant from the standpoint of reducing the rate of violent crime or ameliorating the effects of violent crime in contemporary societies, and to pretend otherwise is both intellectually dishonest and politically dangerous.

14.3.2. Evolutionary Accounts: Phenotypic Plasticity and Developmental Variation

So-called 'evolutionary psychologists' argue that the development of human behavioral tendencies is determined by a number of universal developmental programs that respond to local environmental variation in ways that, in the ancestral environments in which they evolved, were adaptive (see Barkow et al., 1992). In an influential article, Wilson and Daly argued that many behavioral features of 'violent criminals' who grew up in 'high-crime' neighborhoods could be explained as *rational*, adaptive responses to the particular developmental environments in which they found themselves (1997); they argued that we ought to expect, from an evolutionary standpoint, that people growing up in environments with high levels of violence, high levels of economic inequality, and a general lack of 'safe' opportunities for social and/or economic advancement, will become violent risk-takers who strongly discount 'future' prospects in favor of pursuing short-term strategies. This approach is supposed to reveal the 'ultimate' causes of violent behaviors, as well as other behaviors generally considered problematic (see Quinsey, 2002, pp. 1–13). By discovering the 'ultimate' causes, it is hoped that we can both find out more about the root causes of violent crime, and find more effective ways of intervening to prevent it (see Gibbard, 2001; Quinsey, 2002).

While evolutionary psychology, as a field, has come under serious (and in my view, well deserved) attack (see Kaplan, 2002; Buller, 2005), for the purposes of this section, I will not criticize the assumptions made nor the methodological techniques employed. Rather, I will once again stress that *even if* the arguments developed by Wilson and Daly are broadly correct and generalizable to other populations, they in fact have no unique consequences for programs hoping to reduce the prevalence and ameliorate the consequences of violent crime. Rather, a public health approach to violent crime will generally be quite successful concentrating on what Quinsey refers to as the mere 'proximate' causes of violence (Quinsey, 2002); the 'ultimate' evolutionary account, while it would be intellectually interesting if it were better supported, does not in fact have important policy implications.

Wilson and Daly suggest that since our brains evolved to solve 'important' problems regarding maximizing reproductive success within complex social systems, we should expect to find that behavior will vary based on the developmental environment encountered, and that this development should be especially sensitive to the social environment (that is, they hypothesize a form of *phenotypic plasticity* with respect to the social environment of development; see Pigliucci, 2001). In environments in which the average life-span is relatively long, rates of violent death relatively low, opportunities for low-risk social advancement available, and reasonable levels of social success widely available (relatively low levels of social inequality), Wilson and Daly suggest that it would adaptive to adopt a long time horizon (including delaying reproduction), pursue relatively 'low-risk' long-term strategies, and avoid risk-taking (1997, see also Daly and Wilson, 1988). On the other hand, in environments in which the average life span is relatively short and the chance of dying early through violence relatively high, and where there are large social inequalities and low-risk strategies are unlikely to accrue reasonable amounts of social success (so-called 'winner take all' societies), but where risk-taking may have large social rewards, Wilson and Daly argue that it is adaptive to adopt a short time horizon for decision making (including favoring early reproduction), and to pursue high-risk, high-reward strategies, including violent ones (1997).

Wilson and Daly claim that the pattern of life-expectancy, homicide rates, and 'reproductive timing' one finds in neighborhoods in the U.S. conform to this pattern (1997). The neighborhoods with relatively little violence are those with relatively long life-expectancies (and better overall health), relatively little economic inequality (and overall reasonably high incomes), and relatively many educational and economic opportunities; the most violent neighborhoods are those with relatively short life-expectancies (and poor overall health), great economic inequality (and overall very low incomes), and relatively few educational and economic opportunities (1997). Based on this kind of data, authors like Gibbard (2001) suggest that searching for biological differences between more and less violent people may be pointless; the differences between people may not lie in different biology's per se but rather in the different developmental environments to which the people were exposed. While different developmental environments may *produce* different behavioral phenotypes that are related to different brain chemistries, etc., the development of these latter features is *caused* by the environmental differences working on the 'same' developmental systems (Gibbard, 2001; Quinsey, 2002).

While Wilson and Daly note that the likely number of complex 'feedback loops' in the pathways between the developmental environment, reasonable expectations regarding life expectancies, and life-strategies involving violence, etc., are 'daunting' (1997), even if their analysis is right there is no reason to be particularly pessimistic about changes required to reduced violence. If the 'ultimate' causes of violence are the developmental environments encountered, changing those environments ought to reduce violence. While changing the environment may not produce an immediate drop in the rates of violence, changes certainly ought to be effective in the long run. The difficulty that Wilson and Daly seem stuck on, however, is that on their analysis one of the key determinants to becoming a violent adult is growing up in a violent society, and it is hard to see how to change the kind of society that people grow up in without reducing violence through other means.

But of course, violence in societies *can* be reduced by other means; as noted above, there have been some success with programs designed to decrease violence in particular locations, and with programs that aim to increase economic opportunities and reduce social and economic inequalities (see above, and also e.g. Greenwood et al., 1996). A public-health approach to reducing the prevalence and impact of violence in societies would, naturally, focus on those programs that have been shown to reduce violence; interesting, these are same kinds of proposals that attention to Wilson and Daly's work would lead one towards. This is hardly surprising, as the risk-factors associated with violent crime that Wilson and Daly cite are well-known, and the 'proximate' causal pathways between those risk factors and violence are fairly clear. Adding the evolutionary story – the 'ultimate' causes – simply does not add anything to the public policy recommendations that ought to be pursued. So again, while it might be of some intellectual interest to know how human evolution shaped our response to developmental environment (if we could in fact out, see Lewontin, 1998 for a discussion of how difficult this project is), it would *not* have any impact on the policy recommendations emerging from taking a public health approach to violence.

14.3.3. Locally Adapted Populations? Combining Population-Level Genetic Variation and Evolutionary Accounts

(Draper and Harpending, 1982, pp. 255–273) argued that, historically, there were two broad 'reproductive strategies' open to humans; in one, fathers help provide resources for and participate in the rearing of their children, and in other, fathers are broadly absent from the household and contribute little to supporting or raising their children. In the former, they suggest, males will tend to follow long-term, low-risk plans to support their mates and offspring, while in the latter, males will tend to adopt short-term, high-risk strategies to maximize reproductive success; hence, 'father absent' societies will tend to be marked by more competition, more of a 'winner-take-all' style social organization, and far more interpersonal violence than 'father present' societies (Draper and Harpending, 1982, pp. 262–263; see also Harpending and Draper, 1988). Different social arrangements, they argue, favor different 'strategies' – some arrangements favor the 'father absent' strategy, some favor the 'father present' strategy, and some favor a 'mixed' strategy (by favoring the 'rare' strategy) (Draper and Harpending, 1982). For those arrangements that strongly favor one strategy rather than the other, Draper and Harpending suggest that the favored strategy might be maintained by both cultural and genetic inheritance systems (1982 270; Harpending and Draper 1988).

When Draper and Harpending first proposed this model, there was no plausible candidate for a genetic inheritance system to complement the cultural system. With Caspi et al.'s research into the relationship between childhood developmental environment, MAOA levels, and adult violence (see above), one might suppose that a plausible candidate presents itself. In violent 'father absent' societies, low MAOA levels would encourage the development of violent, and hence more 'competitive,' adults; in non-violent 'father present' societies, high MAOA levels would tend to reduce the inter-generational transmission of familial violence when it does occur, and therefore permit more children raised in such circumstances to go on to lead relatively successful lives in the 'father present' model. Mixed strategies, and population polymorphisms, could no doubt be accounted for as well on this model, given certain assumptions about the likely pay-offs of particular alternative strategies in those cultures.

There is no evidence for the existence of such population-level adaptations ('ecotypes' in the biological literature) to different broad reproductive strategies in human populations (for a discussion of the difficulties with finding such adaptations in the human case, see Kaplan, 2002; Pigliucci and Kaplan, 2003). It is *possible* that the different length MAOA promoter regions play such an adaptive role, but evidence for such a claim is currently lacking, and such evidence would be very hard to gather (see Lewontin, 1998). But even if such a claim *could* be well-supported, it would seem to have no implications for policies aimed at reducing the prevalence and ameliorating the impact of violent crime and violence more generally. Again, while such a story, if it turned out to be well-supported, would be of real *intellectual* interest, it wouldn't change what a public health approach to violence prevention would recommend. Once again, attempting to intervene at the level of biochemistry would be inappropriate (see above), and knowledge of the 'ultimate' cause of the prevalence of particular genes and associated behaviors wouldn't be of particular use in intervening.

14.4. CONCLUSIONS: AVOIDING DISTRACTIONS

I have argued that contemporary approaches to finding biological correlates to violence, even if successful, will not prove to be a useful part of public health approaches to preventing violent crime and violence more generally. Of course, I have not discussed every contemporary research project, still less what 'might' someday be discovered. There is a substantial literature on how we ought to deal with particular 'possible' discoveries regarding links between biology and violent crime; various authors have argued that some possible discoveries would force us to rethink our moral intuitions or revise our public policies regarding violent crime (see e.g. Wasserman and Wachbroit, eds, 2001, especially Part II). We are asked to imagine discovering biological correlates to criminal violence that make one 'predisposed' to commit violent acts, or 'impulsive' and thus more unable to 'control' one's temptations than the norm (see e.g. Baron, 2001), or, more dramatically, we are asked to imagine discovering biological pathways that make it *certain* that one will commit violent acts – that *determine* one to become violent (see e.g. Van Inwagen, 2001).

While perhaps interesting as an intellectual exercise, these hypothetical cases should not distract us from what we *do* know about the correlates of violence, and what we do know about how to reduce violence in particular societies. We do not need to wait for more biological data before attempting to reduce the prevalence and impact of violence in our societies, nor is it at all likely that more biological data would even be *helpful* in doing so. What is missing from a public health approach to dealing with violence is not knowledge of the particular biological pathways involved, but the political will to act on what we do know about the (broadly) social pathways involved. Those people concerned with actually reducing the prevalence and impact of violent behavior in societies should therefore focus on the political questions, and treat the biological research as, at best, intellectually interesting distractions from the hard work ahead.

Philosophy Department, Oregon State University, Oregon, USA

REFERENCES

Adam D (2004) Gene scientists plan aggression drug. The guardian (London). 20 July 2004, Home Pages, 3

Barclay, Gordon, Tavares, Siddique (2001) International comparison of criminal justice statistics, 1999. Home office Research and Statistics Directorate, London, Great Britain

Barclay, Gordon, Tavares C (2002) International comparisons of criminal justice statistics 2000. Research Development and Statistics, Home Office U.K. http://www.homeoffice.gov.uk/rds/pdfs2/hosb502.pdf

Barkow JH, Cosmides L, Tooby J (1992) The adapted mind: evolutionary psychology and the generation of culture. Oxford University Press, New York, NY

Baron M (2001) Crime, genes, and responsibility. In: Wasserman D, Wachbroit R (eds) Genetics and criminal behavior. Cambridge University Press, New York, NY

Brunner HG, Nelen M, Breakefield XO, Ropers HH, van Oost BA (1993) Abnormal behavior associated with a point mutation in the structural gene for Monoamine Oxidase A. Science 262:578–580

Buller DJ (2005) Adapting minds: evolutionary psychology and the persistent quest for human nature. MIT Press/Bradford Books, Cambridge, MA

Caspi A, McClay J, Moffitt TE, Mill J, Martin J, Craig IW, Taylor A, Richie P (2002) Role of genotype in the cycle of violence in maltreated children. Science 297:851–854

Cole D (1999) No equal justice: race and class in the American criminal justice system. The New Press, New York, NY

Draper P, and Henry Harpending (1982) Father absence and reproductive strategy: an evolutionary perspective. Journal of anthropological research 38(3):255–273

Day LH (1984) Death from non-war violence: an international comparison. Social science and medicine 19(9):917–927

Daly M, Wilson M (1988) Homicide. Aldine de Gruyter. Hawthorn, NY

DiLalla, Fisher L, Gottesman II (1991) Biological and genetic contributors to violence – widom's untold tale. Psychological bulletin 109(1):125–129

Dussich J, Fujiwara Y, Sagisaka A (1996) Decisions not to report sexual assault in Japan. In: Sumner C, Israel M, O'Connell M, Sarre R (eds) AIC Conference Proceedings No. 27: International victimology: selected papers from the 8th International Symposium: proceedings of a symposium, Australian Institute of Criminology, Canberra, 21–26 August 1994

Elliott DS (1998) Life-threatening violence is primarily a crime problem: a focus on prevention. University of Colorado law review 69(4):1081–1098

Freudenberg N (2000) Health promotion in the city: a review of current practice and future prospects in the United States. Annual review of public health 21:473–503

Gibbard A (2001) Genetic plans, genetic differences, and violence: some chief possibilities. In: Wasserman D, Wachbroit R (eds) Genetics and criminal behavior. Cambridge University Press, Cambridge, UK

Gibbs WW (1995) Seeking the criminal element. Scientific American. March 1995, pp 101–107

Gould SJ (1981) The mismeasure of man. W.W. Norton & Company, New York, NY

Greenwood PW, Model KE, Rydell CP, Chiesa J (1996) Diverting children from a life of crime: measuring costs and benefits. RAND, Santa Monica, CA

Guerrero R (2002) Violence is a health issue. Bulletin of the World Health Organization 80(10):767

Hacking I (2001) Degeneracy, criminal behavior, and looping. In: Wasserman D, Wachbroit R (eds) Genetics and criminal behavior. Cambridge University Press, Cambridge, UK

Hamer D, Copeland P (1998) Living with our genes: why they matter more than you think. Doubleday, New York, NY

Harpending H, Draper P (1988) Antisocial behavior and the other side of cultural evolution. In: Moffitt TE, Mednick SA, Nijhoff M (eds) Biological contributions to crime causation. Dordrecht, Netherlands

Hawkes Nigel (2002) Gene that may turn abused men to violence. The Times (London). 2 August 2002, Home News, p 8

Kaplan J (2000) The limits and lies of human genetic research: dangers for social policy. Routledge, New York, NY

Kaplan J (2002) Historical evidence and human adaptations. Philosophy of science 69(3S): S294–S304

Kellermann AL, Fuqua-Whitley DS, Rivara FP, Mercy J (1998) Preventing youth violence: what works? Annual review of public health 19:271–292

Kennedy BP, Kawachi I, Prothrow-Stith D (1996) Income distribution and mortality: cross sectional ecological study of the Robin Hood index in the United States. British medical journal 312:1004–1007

Kennedy BP, Kawachi I, Prothrow-Stith D (1998) Income distribution, socioeconomic status, and self rated health in the United States: multilevel analysis. British medical journal 317:917–921

Kevles DJ (1985) In the name of eugenics: genetics and the uses of human heredity. University of California Press, Berkeley, CA

Koenen KC, Moffitt TE, Caspi A, Taylor A, Purcell S (2003) Domestic violence is associated with environmental suppression of IQ in young children. Development and psychopathology 15:297–311

Krug E, Mercy J, Dahlberg L, Zwi A (2002) The world report on violence and health. The Lancet 360(9339):1083–1088

Krug EG, Powell KE, Dahlberg LL (1998) Firearm-related deaths in the United States and 35 other high- and upper-middle-income countries. Int J Epidemiol 27:214–221

Krug EG, Sharma GK, Lozano R (2000) The global burden of injuries. Am J public health 90(4):523–526

Lester D (1986) The distribution of sex and age among victims of homicide: a cross-national study. Int J Soc Psychiatry 32(2):47–50

Lester D (1991) Crime as opportunity. Brit J criminol 31(2):186–188

Lewontin RC (1998) The evolution of cognition: questions we will never answer. In: Scarborough D, Sternberg S (eds). Methods, models, and conceptual issues: an invitation to cognitive science, vol 4. The MIT Press, Cambridge, MA

Lewontin RC, Rose S, Kamin LJ (1984) Not in our genes: biology, ideology, and human nature. Random House Inc., New York, NY

Lowenstein LF (2002) The genetic aspect of criminality. Justice of the peace 166:767–772

Macdonald G (2002) Violence and health: the ultimate public health challenge. Health promotion international 17(4):293–295

Mann CC (1994) Behavioral genetics in transition. Science 264:1686–1689

Martens WHJ (2002) Criminality and moral dysfunctions: neurological, biochemical, and genetic dimensions. Int J Offender Ther 46(2):170–182

McCulloch A (2001) Social environments and health: cross sectional national survey. BMJ 323:208–209

McDonald D (2000) Violence as a public health issue. Trends and issues in crime and criminal justice 163:1–6

Mercy JA, Krug EG, Dahlberg LL, Zwi AB (2003) Violence and health: the United States in a global perspective. Am J Public Health 92(12):256–261

Moffitt TE, Mednick SA (1988) Biological contributions to crime causation. NATO Advanced Study Institute, Series D: Behavioral and Social Science, Number 40. Martinus Nijhoff Publishers, Dordrecht, The Netherlands

Mugford J, Nelson D (eds) (1996) Violence prevention in practice: Australian award-winning programs. Australian Institute of Criminology

Pigliucci M (2001) Phenotypic plasticity: beyond nature and nurture. John Hopkins University Press, Baltimore, MA

Pigliucci M, Kaplan J (2003) On the concept of biological race and its applicability to humans. Philosophy of Science 70(5):1161–1172

Quinsey VL (2002) Evolutionary theory and criminal behaviour. Legal and criminological psychology 7(1):1–13

Rawson B (2002) Aiming for prevention: medical and public health approaches to small arms, gun violence, and injury. Croatian medical journal 43(4):379–383

Rensberger B (1992) Science and sensitivity: primates, politics, and the sudden debate over the origins of human violence. The Washington post. 1 March 1992. Outlook, C3

Rockett IRH (1998) Injury and violence: a public health perspective. Population Bulletin

Rodgers GB (2002) Income and inequality as determinants of mortality: an international cross-section analysis. Int J Epidemiol 31:533–538

Sarkar S, Fuller T (2003) Generalized norms of reaction for ecological developmental biology. Evolution & development 5:106–115

Schaffner KF (2001) Genetic explanations of behavior: of worms, flies, and men. In: Wasserman D, Wachbroit R (eds) Genetics and criminal behavior. Cambridge University Press, Cambridge, UK

Sen A (1999) Development as freedom. Random House, New York, NY

Sober E (2001) Separating nature and nurture. In: Wasserman D, Wachbroit R (eds) Genetics and criminal behavior. Cambridge University Press, Cambridge, UK

Taylor KA (2001) On the explanatory limits of behavioral genetics. In: Wasserman D, Wachbroit R (eds) Genetics and criminal behavior. Cambridge University Press, Cambridge, UK

The Home Office Research Department. Criminal justice: the way ahead. CM 5074. http://www.archive.official-documents.co.uk/document/cm50/5074/5074.htm Published by the Stationary Office, UK

Van Inwagen P (2001) Genes, statistics, and desert. In: Wasserman Dand R. Wachbroit (eds) Genetics and criminal behavior. Cambridge University Press, Cambridge, UK

Wasserman D, Wachbroit R (eds) (2001) Genetics and criminal behavior. Cambridge University Press, New York, NY

Wilson JQ (1994) What to do about crime. Commentary 98(3):25–35

Wilson M, Daly M (1997) Life expectancy, economic inequality, homicide, and reproductive timing in Chicago neighbourhoods. BMJ 314(8089):1271–1274

Winkelmann J (2003) MAOA: susceptibility locus for the severity of RLS phenotype? Sleep Medicine 4:81–82

15. TAKING EQUIPOISE SERIOUSLY: THE FAILURE OF CLINICAL OR COMMUNITY EQUIPOISE TO RESOLVE THE ETHICAL DILEMMAS IN RANDOMIZED CLINICAL TRIALS

15.1. INTRODUCTION

The general concept of equipoise may be defined as uncertainty about or a lack of a grounded preference between two arms of a clinical trial (Fried, 1974; Hellman and Hellman, 1991). This is often said to specify the circumstances under which an RCT, in which subjects are randomized to either of two treatments, is morally legitimate. For if there is no grounded preference for one treatment, randomization to either will not generate a moral tension. This general concept of equipoise cannot resolve the RCT dilemma. One reason for this is that, as data accumulate, equipoise will be disturbed long before we obtain the statistically significant results that was the point of the trial.

In this essay I provide a critical analysis of one especially influential attempt to resolve this problem, the clinical equipoise standard proposed by Benjamin Freedman in 1987 in the *New England Journal of Medicine* (Freedman, 1987). Freedman claims that if we conceptualize the matter in terms of what he terms 'clinical equipoise', instead of what we have implicitly been focusing on (what he terms 'theoretical equipoise'), then we will resolve the basic dilemma posed by RCTs.

This notion has had a lot of influence (Freedman, 1990, 1992; Crouch and Arras, 1998). Indeed, a pair of articles in the *British Medical Journal*'s 'Education and Debate' section in 2000 debates the merits of clinical equipoise and the 'uncertainty principle' and gives the impression that these concepts count as the preferred justifications in North America and the U.K. respectively (Weijer et al., 2000).

A careful analysis reveals that this proposal is simply mistaken. We cannot honestly tell ourselves that we avoid, resolve or even properly address the moral dilemma that RCTs pose for us by relying on the claim that we are in this state that Freedman calls 'clinical equipoise'.

Indeed, I submit that this criterion's appeal comes in part from its not being one criterion at all, but rather various different criteria that get conflated. The most central pair of concepts that we need to disentangle here is that of clinical equipoise properly so-called, and community equipoise. While a textual basis for each can be found in Freedman's article, they in fact represent two quite different ways of moving away from the original equipoise concept. In addition, within each of clinical and community equipoise, there is further ambiguity.

It is important to see that these ambiguities do not just make the concept a little fuzzy, generating a bit of vagueness at the edges, something that could be tidied

H. Kincaid and J. McKitrick (eds.), Establishing Medical Reality, 215–233.

up into a clear and clearly justified principle. Instead, they mislead us into thinking that there is some one notion that captures a number of important features, when in fact it is an incoherent hodgepodge of concepts. Freedman's use of the term 'clinical equipoise' applies to a set of distinct concepts that in fact would provide distinct and incompatible guidance, and hence between which we need to choose. When one analyzes these specific concepts individually, one sees that neither of them provides a justification nor an adequate ethical guide for RCTs.

It is very important whether we conceptualize the ethical issues in RCTs in terms of 'clinical equipoise' or in some other way, so it is worth examining with further scrutiny. Indeed, a careful evaluation of this matter is fundamental for proper judgments about when to start and stop a clinical trial, to evaluating the ethical issues surrounding informed consent and patient communication that arises within them, and in considering the ethics of human subject research generally. Further, these debates bring us in contact with a number of important topics in the philosophy of medicine, such as clinical judgment and its relationship to group consensus and other social processes in the creation and establishment of medical knowledge.

15.2. THE RCT DILEMMA AND THE INTERIM DATA PROBLEM

The traditional view is that an RCT is morally legitimate just in case there obtains a state of equipoise about the relative efficacy of the two treatment arms of a study (or about the hypothesis that one arm is better than the other). For then random assignment of one's patient cannot be seen as provision of an inferior alternative.

This generic notion of equipoise does arguably yield a sufficient standard; the presence of equipoise would seem to provide a reasonable justification for allowing randomization of the patient and good guidance in judging under what circumstances an RCT would be morally justified.

The skeptic will suggest that this condition is in fact rarely satisfied, on the grounds that there will typically be some reasons tilting one way or the other due to background knowledge or preliminary evidence (for instance, from phase II trials). This skepticism is made stronger and more broadly applicable by focusing on the problem of accumulating interim data. For even if a trial *begins* at complete equipoise, once there is a trend in one direction which is sizable but does not yet reach the pre-determined level of statistical significance, equipoise will be lost.

Consider the situation where interim data support the claim 'Treatment A is better than treatment B' as justified at, say, the .07 level of statistical significance – meaning that there is no more than a .07 probability that the data would show this difference if there were in fact no differences between the treatments. It is hard to say that this evidence is completely irrelevant to the question of how to act on a patient's behalf. Granted the conflict is *less* strong than it would be at the .05 level (let us suppose this is the point at which it has been decided by the study's designers that the trial will be complete), but such data still provide evidence on which it would be rational to act.[1]

There is a tendency to deny this moral conflict by saying that in such a situation 'we really don't know' which treatment is better (and hence that it is indeed not rational to act on it). For in order not to be fooled by such trends in the data, we have taken up the policy specifying the .05 level as what counts as legitimate scientific evidence, and we know of cases where acting on less than this has led us astray. This is a mistaken line of reasoning. It presupposes an overly simple conception of the nature of knowledge – or, we should say, of our confidence about that knowledge. It assumes that such knowledge, or confidence, pops into being all at once, while in fact it comes in degrees. This fact forces upon us the question of determining how strong that evidence has to be before we accept the hypothesis at issue.

The idea is that all trials where we accumulate evidence along the way will pass through an uncomfortable period during which two things will happen. (1) We will no longer be in equipoise (there will be some evidence pointing in one direction) and (2) we will not yet have enough evidence to stop the trial on grounds of having obtained the evidence that the trial was aiming for. Thus carrying out RCTs all the way to statistical significance (the point required in order to be doing the science responsibly) requires continued randomization of patients when there is *some* reason to believe that A is better than B (i.e., past the point of equipoise).

The key things to stress here are, first, that security of knowledge comes in degrees, and second, that we must distinguish the following two sorts of *decision contexts*: The *present patient decision* concerns whether the evidence warrants performing the single act of choosing A over B for one's present patient. The *policy decision* concerns whether the evidence warrants terminating the trial and deciding to recommend A over B as a general policy from now on. The crucial point is that the decisions made in these different contexts have different things at stake, and therefore require different degrees of security of knowledge and hence different amounts of evidence.[2] The individual patient decision to treat the patient with A has serious but limited consequences: if it turns out that you are mistaken and that B is actually the better drug, the consequences are that this patient gets the wrong drug. There will be another decision down the road for the next patient, and the next, and we can make *those* decisions once we get more evidence. (If the mistake involves only that A is *no better than* B, the consequences are even less serious.)

But the consequences of being wrong (mistakenly accepting the hypothesis) in relation to the policy decision are more severe (e.g., approval and marketing of an unsafe or ineffective drug, affecting more people than the present patient, or foreclosing the possibility of gaining further controlled data.). As a result, it is appropriate to require a greater amount of evidence before accepting (or acting on) the conclusion.

A number of different responses can be made to this conclusion that RCTs carried out to statistical significance will violate the ethical standard of equipoise. These run from the claim that such trials are justified on utilitarian grounds (a great good will come about for many future patients at the cost of a certain

amount of suboptimal treatment for limited number of present research subjects),
or because the subjects have given informed consent, to the fact that the clinicians
do not themselves fall out of equipoise because the interim data are kept secret by
data and safety monitoring boards (Marquis, 1983; Gifford, 1986). In this paper,
we focus our attention only on Freedman's proposed justification – that, even
if we see the accumulating data, we don't really fall out of the relevant sort of
equipoise, 'clinical equipoise'.

15.3. FREEDMAN'S PROPOSAL

Freedman points out that equipoise has traditionally been understood in an individual
manner, as the requirement that the individual investigator be in a state of genuine
uncertainty concerning the relative efficacy of treatments A and B. As such, he
claims, it is conceptualized in a way that makes it both subject to irrelevant features
of the situation and, at the same time, too easy to topple (being balanced on a
knife-edge).

Freedman proposes instead that the correct or morally relevant standard to be
concerned about is that which he calls *'clinical* equipoise'. Rather than requiring that
the individual investigator be in this state of genuine uncertainty, it is enough that
there be genuine uncertainty within the medical community about the comparative
merits of the two treatments. He elaborates that such equipoise is said to obtain when
there is 'present or imminent controversy in the clinical community' (Freedman,
1987, p. 141). He also states that equipoise is disturbed precisely '(a)t the point
when the accumulating evidence in favor of B is so strong that the committee
of investigators believe no open-minded clinician informed of the interim results
would still favor A' (Freedman, 1987, p. 144).

Thus even though the physician may personally prefer treatment A, patients can
legitimately be entered into a randomized trial so long as there remains disagreement
in the community of medical professionals or experts.

This is said to avoid the problems of extreme fragility associated with theoretical
equipoise. It is also suggested that it has more clinical relevance, apparently taking
into account the sort of judgments that can and ought to be made in the real,
clinical setting. For instance, it is said to focus on questions of clinical as opposed
to theoretical significance, and takes into account idiosyncrasy and sensitivity to
various arbitrary modes of representation(1987, p. 413).

Freedman holds not only that this works as a criterion to specify under which
circumstances trials would be justified, but also that this will capture our present
good practice, and hence justify roughly the practice that IRBs carry out now in
their review of research protocols (1987, p. 145). Importantly, he holds that it shows
that we can justify this general practice without having to appeal to some utilitarian
justification.

There are a number of suggestive and important ideas worthy of further reflection.
But in fact I will show that such reflection raises skepticism about whether we
are presented here with a solution to the dilemma. Indeed, part of the problem is

that there are so many different things happening at once. There is more than one criterion being put forward under the same name. An important part of my strategy will be to discern critical ambiguities that plague the proposal.

15.4. COMMUNITY EQUIPOISE VS. CLINICAL EQUIPOISE

The central ambiguity is whether the proposal is for community equipoise or clinical equipoise. Freedman terms his concept 'clinical equipoise', and when he describes the old, traditional, misleading concept that led us astray, he terms it 'theoretical equipoise'. But if we look at most of what is said above, and especially that which is stated as a (relatively) explicit criterion, surely what we see is a proposal that we follow *community* equipoise (replacing the traditional standard of *individual* equipoise). It's about all the various judgers coming to the same conclusion, and it's not clear that there is something specifically clinical about it. (Of course, the community referred to is the clinical community, but clinical is not what is new here, as the individual who is said to fall out of equipoise too early was, of course, a clinician.) On the other hand, some of his points do in fact concern some notion of clinical vs. theoretical. So it seems that there are two different proposals here. We need to consider them one at a time.

Consider first the individual/community contrast:

INDIVIDUAL EQUIPOISE

It is morally acceptable to enter a patient into a trial *precisely insofar as*:
- you as an individual (clinician) are indifferent between (i.e., genuinely uncertain about the relative merits of) treatments A and B. (For instance, this is sometimes described as the claim that you would be willing to choose randomization between A and B for yourself or a loved one.)

COMMUNITY EQUIPOISE

It is morally acceptable to enter a patient into a trial (even if you have a preference – i.e., even if you are out of individual equipoise) so long as:
- the 'clinical community' is genuinely uncertain
- i.e., there is not (yet) a consensus within the clinical community about the comparative merits of A and B

Community equipoise is a less strict (more easily satisfied) requirement than individual equipoise, and so this will allow a longer period of time during which it is morally acceptable to continue to give patients any of those treatments on a randomized basis. This would of course allow more research to be done. Despite the label Freedman gives to his criterion and the title of his paper, community equipoise in this sense seems to be the central thrust of his proposal.

But other statements referred to above suggest a distinct contrast – concerning what might truly be described as clinical equipoise (in contrast with the traditional notion of theoretical equipoise). What this involves is harder to say. I believe that the best analysis explains this in terms of the type of question that we are trying to answer:

Clinical equipoise is equipoise about a practical question aimed at clinical practice and the goal of healing, such as 'Is A better treatment for treating these patients than B is?' *Theoretical equipoise*, on the other hand, is equipoise about a theoretical question aimed at general scientific knowledge and the goal of understanding, such as 'Does treatment A cause outcome O with greater frequency (and extent) than does B?' (This characterization fits with Freedman's application of this concept to his analysis of placebos in (1990).)

The clinical question is of course claimed to be the ultimately important one – being not merely theoretical, but tied to particular actions that will be taken, and tied to clinical reality.

We will explore later in the paper whether clinical equipoise in this sense can address the RCT dilemma. But let me first emphasize again that clinical equipoise and community equipoise are distinct notions, and hence, even if it were to be establish clinical equipoise as the appropriate conceptualization, this would say nothing about whether community equipoise is to be seen as the right standard. For it is one thing to distinguish two kinds of questions, a theoretical question about whether a drug has a causal effect on the incidence of a certain simple, well-defined outcome, and a practical or clinical question about whether that drug is a better treatment overall for a certain set of patients than is another drug. But it is a different matter entirely to distinguish two modes or standards of assessment of these questions: 'What do I think concerning whether there is evidence for the claim?' or, 'Is there community agreement concerning this?' Note that we could apply each of these standards to either kind of question.

Freedman's presentation also mentions other elements that do not fall neatly into either of these types; some of these will be discussed briefly later in the paper. My main point at present is that the presentation given by Freedman and others suggests that this is all one concept, but it is not. We cannot evaluate the proposal reasonably without understanding this. In particular, we must guard against the temptation, upon finding some positive feature concerning, say, clinical (as opposed to theoretical) equipoise, to then view this as counting in favor of community equipoise. The worry is that Freedman's 'clinical equipoise' concept looks more plausible than it really is precisely because one does not look carefully enough at it.

In any case, I shall argue that none of these criteria (or senses of equipoise) has the resources to enable us to really provide a resolution to the ethical dilemma posed by RCTs.

15.4.1. Evaluation of Community Equipoise

First, we will look at the claim that *community* equipoise is to be preferred over individual equipoise (leaving aside the clinical/theoretical distinction). There will be two components to my critique. First, I challenge the claim that it is *morally acceptable to follow* the community equipoise standard – questioning whether the policy of deferring to the fact of disagreement is morally legitimate, and arguing that such extension of trials is indeed done at the expense of our obligation to do the best for our patients. Since following the principle is not actually legitimate, it does not resolve our moral problem to do so.

Secondly, I argue that, even if we were to grant that following community equipoise was justified (that one could follow it confident that this did not involve treating subjects unethically), this still wouldn't solve our problem. For even though adherence to the community equipoise criterion will allow *more* trials (or allow trials to go *longer*), it is still not sufficient to allow us to extend trials long enough to obtain the requisite security of knowledge. Community equipoise would, at least typically, be disturbed long before we attain the traditional criterion of statistical significance – or before we attain any other reasonable criterion for having gained an appropriate level of security of knowledge – which was the point of the trial. If we were only to carry out trials to this point, they would not be worth doing. (Cf. Gifford, 1995.)

15.4.2. Challenging the Claim that following Community Equipoise is Morally Acceptable

In order to appreciate the nature of the community equipoise thesis, it is important to consider the following rather striking feature: it shifts our attention from numerical values of statistical significance, or the question of whether evidence exists, to a counting up of people who hold certain opinions. It might be said to shift us from the evidential or epistemological point of view to the *sociological* point of view (on the face of it, from something objective to something subjective). This is not a problem *per se*. It would indeed seem to capture the important insight that science (and hence medical science) is a social process. Still, the shift from evidential to sociological perspective is a quite significant fact that should be kept in mind.

A way in which this could in fact be argued to be a liability is suggested by the following: Use of this criterion involves deferring to the fact of disagreement by others, and thus involves setting aside personal individual judgment. Surely, one's individual clinical judgment should carry some moral weight. Consider: I have evidence that seems *to me* on reflection to warrant choosing one way (if I had to choose). If it were me or a loved one I'd use the information. The Principle of Community Equipoise says that, with respect to some patient who may be involved in a clinical trial, one's own judgment is to be overridden. There is of course reason to question or be uncertain about one's own judgment, but there are also reasons to question or be uncertain about the judgments of others. Note well that it will not do to say that it is being overridden in order to advance the greater good of clinical research. The community equipoise criterion is supposed to function in a way that avoids allowing the greater good to be determinative in this way.

Add to this a more explicit argument from the autonomy and integrity of the individual clinical practitioner: one has the right and duty to think for oneself in deciding what is best for their patient. One way of worrying about the 'sociologicization' idea is that it makes an individual merely an unthinking conduit of the opinions of others in the medical community.[3]

In light of all this, the proponent of the community standard needs to be able to provide some serious arguments on the other side. The most important rationales in support of the community equipoise criterion would appear to be the following: First is the 'evidential warrant' rationale, that the judgments of others count as evidence and thus should modify one's opinion. Second is the 'no worse off than otherwise' rationale: subjects entered into a trial operating according to community equipoise would not have done any better outside of the trial.

As stated above, my analysis here will make use of the strategy of uncovering ambiguities in the community equipoise criterion. The criterion can be understood in more than one way, and these different ways are in conflict with one another. Understanding it one way has certain advantages; understanding it another way has others. Again, I insist that one needs to decide between these things, and I also argue that no single interpretation emerges as attractive.

15.4.3. Ambiguities

The first ambiguity concerns what is to count as the community, or as the set of individuals whose opinions will count in constructing the community opinion. The question actually has a number of components: How large a group is this to be? According to what criteria will they be chosen? How 'expert' will they have to be (how marginal can one be and still be part of the community)? While it might be fruitful to explore several of these aspects, I will here simply make a distinction between two idealized contrasting interpretations: 'narrow' and 'broad'. We thus have the following two quite different criteria:

COMMUNITY EQUIPOISE OF THE *NARROW* COMMUNITY

It is morally acceptable to enter a patient into a trial (even if you have a preference) so long as
* the narrowly defined set of experts is genuinely uncertain
* i.e., there is not (yet) a consensus amongst the narrow set of experts about the comparative merits of A and B

COMMUNITY EQUIPOISE OF THE *BROAD* COMMUNITY

It is morally acceptable to enter a patient into a trial (even if you have a preference) so long as
* the broadly defined medical community is genuinely uncertain
* i.e., there is not (yet) a consensus of the whole medical community (e.g., the set of physicians making treatment decisions about this class of patients) about the comparative merits of A and B

The main point here is that the narrow and broad interpretations of community equipoise provide distinct standards. The rationales that would justify these two standards will be different. (As will be seen, the narrow interpretation resonates with the 'evidential warrant' rationale discussed below, and similarly, the broad interpretation resonates with the 'no worse off than otherwise' rationale.) And the length of time they would allow RCTs to continue would be different. Clearly the broad community standard would justify continuing longer, for there will in all likelihood be some members of the broader community still undecided (or still holding 'the old view') some time after all those in the narrow community have all accepted the change.

The second ambiguity concerning the community equipoise standard involves the question of how the community judgment is to be composed or generated out of these individual judgments. Are we to require unanimity, or only what might be called the 'preponderance of opinion'? This suggests a *different* contrast between two criteria:

UNANIMOUS COMMUNITY EQUIPOISE (FREEDMAN)

It is morally acceptable to enter a patient into a trial (even if you have a preference) so long as
- the community is genuinely uncertain in the sense that not all individuals judge equipoise to be disturbed (in the same direction)
- i.e., there is not (yet) a unanimous consensus in the community favoring A over B

On this view, equipoise is disturbed only when the last hold-out gives up.

'PREPONDERANCE OF OPINION' COMMUNITY EQUIPOISE

It is morally acceptable to enter a patient into a trial (even if you have a preference) so long as
- the community is genuinely uncertain in the sense that there is not a majority of individuals judging equipoise to be disturbed in the same direction
- i.e., there is not (yet) a 'preponderance of opinion' favoring A over B

Again, we must emphasize that the unanimity and preponderance interpretations of community equipoise constitute distinct standards. Hence the rationales that would justify them will be different, and the length of time they would allow RCTs to continue would be different. In this case, clearly the unanimity standard is a harder one to disturb and hence would justify continuing longer. Conversely, the 'preponderance' standard could break down quite quickly (indeed, possibly even before one's individual equipoise does). Clearly Freedman needs to hold the unanimity view (or, conceivably, something in between, much closer to unanimity).

These two distinctions generate for us a 2x2 matrix of potential sorts of community equipoise: unanimous narrow community equipoise, unanimous broad community equipoise, 'preponderance of opinion' narrow community equipoise, and 'preponderance of opinion' broad community equipoise.[4] While I will say some things about which interpretations appear more appropriate, my more specific goal here is simply to suggest that there is no one obvious choice, and that these subtleties were not thought through in the original proposal or in subsequent commentaries. Indeed, I think that the perceived plausibility of Freedman's clinical equipoise view rests on leaving these ambiguities intact. Teasing them apart shows that no single view remains attractive.

Having now described these ambiguities as background, let us now consider the candidate rationales for justifying community equipoise.

15.5. THE EVIDENTIAL WARRANT RATIONALE

The 'evidential warrant' rationale, that the judgments of others count as evidence and hence should cause one to change one's opinion, may sound compelling. Clinician should take into account the fact of disagreement with others when trying to decide what to believe. For example, clearly you should take this as a reason to go back and look harder at your evidence and your reasoning.

These facts do not establish the community equipoise criterion, because community equipoise asserts more than that. It says that you should act on those other views – mold your behavior on the basis of those other beliefs – regardless of whether or not on reflection you accept their reasons.

Still, there are some reasons to take this seriously. First is one's knowledge that one is often wrong in one's judgment (even one's judgment on reflection). It might be argued that the community equipoise criterion is an appropriate means to try to keep these things in check.

Another reason is this. Suppose you hear that a number of physicians from a wide variety of backgrounds have all (independently) come to accept that A is better than B. This might be argued to be indicative of a certain robustness of the evidence for this claim. Importantly, it can make sense to view the fact that these others have these opinions as evidence (independent of whether you accept their reasons), not just as a heuristic motivating a closer look. We might analogize these judgments by others to probes you are able to put out when you cannot check on such matters for yourself. The fact that some other scientist confronted with this evidential input comes to a certain conclusion is itself evidence of something – the more so the more positive things you know about their training and track-record. Physicians (and scientists) do in fact rely on the views of others in this way all the time. This is one of the ways in which science is a social process.

Note that since it is those experts that make up the narrow (expert) community that have the most claim on us to follow their judgment (even without agreeing with or understanding their reasons), the evidential warrant rationale suggests strongly the narrow as opposed to the broad interpretation of community equipoise.

Whatever rationale there is of the sort we have just examined (that is, even accepting that there is reason to modify your belief directly, based on the bare fact of the disagreement of others), it is hard to see why this will take you all the way to the unanimity version of community equipoise, all the way until the last person comes on board. This is especially implausible if this is taken in the *broad* sense (where we consider the whole medical community), but the problem exists also in the case of the *narrow* (expert) view.

To see this, let us grant that it is rational to be swayed by the views of others on grounds that their views really count as evidence. The reasoning that makes this make sense argues strongly in favor of the preponderance of opinion view over the unanimity view.

Suppose, for instance, that, amongst the expert community, we have reached the point where the ratio of those who prefer A to those who prefer B is 4:1, 9:1, or even 99:1. It is clearest when the skew is very great, but even in other cases; it seems rational to take this to be evidence in favor of the majority view. That is, if we are going to be swayed by the judgment of others in the way that community equipoise proposes, and if the rationale for this is this evidential one, then we should find it reasonable to be swayed by others in this very specific way – according to the majority opinion (as long as it's not close to evenly split), not in the more arbitrary way of whether the vote is unanimous or not. For consider: You think A is better. 4 out of 5 experts agree. This should give you a reason to modify your belief in the other direction.

In light of this, it is not clear what reason one could have to adopt the unanimity view, except perhaps the desire to have a more conservative criterion, which will allow more testing to occur. However, this would of course beg the question. This therefore suggests the preponderance view. But of course this preponderance version of community equipoise is much more easily disturbed, and so we will not be able to carry on the trials very far at all.

Thus, I conclude that the Evidential Warrant rationale will not justify reliance on a community equipoise that will solve the problem of fragility had by Individual Equipoise. What plausibility there is to community equipoise is had only by the more fragile 'preponderance' view.

15.6. THE 'NO WORSE OFF THAN OTHERWISE' RATIONALE

The other rationale that we said should be taken seriously is the 'no worse off than otherwise' rationale. Subjects entered into a trial operating according to community equipoise would not have done any better outside of the trial, so their treatment is not suboptimal in that sense. For suppose the medical community is divided; any given patient might on their own go to a physician favoring A, or they might go to a physician favoring B. Patients in effect randomize themselves. So joining the trial, relative to their actual alternative does not worsen their situation.[5]

This surely has some force. Of course, there remains the fact that your judgment is that A is better, that you think that doing your best would require giving A. The 'Otherwise' rationale seems not even to address this, highlighting that what we have here really is a standard concerning what patients have a right to, rather than what would be best for them.[6]

Indeed, notice how this is a substantially different rationale: In the case of the evidential warrant rationale, the claim is that we really can't tell, or cannot give objective reasons about, which is better. But here, under the 'no worse off than otherwise' rationale, we have a lowering of the standard: 'You could have gotten what I think is better, but the CE criterion says that there is no moral requirement to provide this. It might be argued that such lowering is justifiable, perhaps analogously to the claim that finite budgets require that we ration care and only allow treatments with a certain level of costworthiness. But if so, this fact needs to be faced squarely, and it will have important implications for such matters as how the situation ought to be presented in the informed consent process.

But note also that the story told above about patients randomizing themselves throughout the medical community definitely suggests the *broad* sense of community (the community of practicing physicians), rather than the *narrow* sense of community experts (which is what the Evidential Warrant rationale was tied to). It is to someone in the broader community that they would have gone otherwise. This raises a question of whether the two different rationales will work together or at cross-purposes.

Further, a closer look at what really would have happened to the subjects otherwise reveals another awkwardness about this justification. First, we tend to have in mind here the situation where a new, experimental drug is gradually shown to be better. But note that this Otherwise rationale (for continuing to randomize some to the less favored treatment) will not work so well if the 'trend' is against the experimental drug, where the patient receiving the drug within the trial would otherwise have been getting the standard drug outside the trial.

And second, suppose the ratio of doctors who prefer A to those who prefer B is 4:1, and hence these are the odds of getting the drug presumed better if one were outside the trial. We can really only justify the claim that the patient would be no better off outside the trial if we randomize between the two arms in a 4:1 ratio. This would be possible to do, but it is not the intention of those who propose community equipoise, and it would require a longer trial and subject more patients to it.

So I conclude that the Otherwise rationale likewise is problematic from the moral point of view, and in any case it pulls in a different direction than the Evidential Warrant rationale. The Otherwise rationale would appear to support the broad interpretation, while the Evidential Warrant rationale supports the narrow, preponderance view. Thus, overall, the case for it being morally acceptable to follow community equipoise is weak.

15.7. THE ARGUMENT THAT COMMUNITY EQUIPOISE WON'T GET US THE WHOLE WAY ANYWAY

Suppose that despite what has been argued above, these rationales are successful and justify the same policy. Suppose, that is, that we have decided that it is perfectly morally acceptable to follow community equipoise (even the unanimity version, leaving aside the argument that the weaker, preponderance of opinion, standard seems to be the more plausible interpretation). Even then, how far would this take us? I claim that it quite clearly would not take us far enough to obtain the degree of security of medical knowledge, which was the goal in the first place. Community equipoise would at least typically be disturbed long before the traditional statistical significance criterion (or any other reasonable criterion for having gained an appropriate level of security of knowledge) is achieved. Hence even if following community equipoise was morally acceptable, it would not resolve the RCT dilemma as originally posed.

The way to see this is to note the flaw in Freedman's reasoning about this. He asserts that following community equipoise will indeed be sufficient (and will justify roughly our present practices). His reason is that community equipoise is defined in such a way that it is disturbed when there is no longer disagreement in the community about the relative merits of the two arms, which is just what it is to resolve the question, which was the point of the trial (1987, p. 144).

But this reasoning ignores the problem with which we began, implicitly confusing the distinction presented at the outset between *present-patient* and *policy* decisions. It presumes that when the last person changes her mind and says 'OK, there is now enough evidence to convince me to give A', she changes her mind simultaneously with respect to both the present patient decision and the policy decision. But this would be inappropriate. A greater level of confidence is necessary for the policy decision; the consequences of accepting the hypothesis when it is wrong are more serious.[7]

When the last person comes on board (falls out of individual equipoise) with respect to the present-patient question, this should be because she has just barely enough information to be convinced of this – just enough data to warrant making the choice for the one patient. Therefore she should *not* be ready to come on board with respect to the *policy* question and stop the trial on grounds that we have attained the scientific answers. For the latter requires more evidence than does the former. So even once community equipoise is disturbed (that is, even once the last individual crosses this line), there will be good reasons *not* to stop the trial and take the policy action on the basis of the trial information up to that point. After all, someone (this last holdout) whose judgment up until this point you respect enough to use to warrant the claim:

'we can keep putting people in the trial, because there's no *real* evidence that the arms are unequal' (or there is uncertainty about that claim)

is now saying

'I judge that there is not yet enough evidence to warrant making the *policy* judgment'.

Could it be argued that she is not being reasonable? What would you say to her? That she should change her view (that is, strengthen her conviction to the policy-relevant level) on grounds of the 'preponderance'? Clearly it would be hypocritical to espouse this line unless we were consistent and had used the preponderance view before. (But then community equipoise would have been disturbed long ago.) In any case, there may in fact not yet be preponderance. Nothing guarantees – or really even suggests – that half of the judges have crossed the policy line. Indeed, it doesn't follow from all of the judges having crossed the first (present-patient) line that *any* of them have crossed the second (policy) line. It depends on how broad the spread of views is amongst the community members.

I conclude that following community equipoise would not in fact capture our present good practice in the way that it might initially seem to.

15.8. CLINICAL VS. THEORETICAL EQUIPOISE

As mentioned above, not all of what Freedman discusses falls under this concept of *community* equipoise. And some of these other features are more plausibly construed as fitting his term 'clinical' (as opposed to theoretical) equipoise.

Now, these other features do not themselves all fall neatly together under one criterion, and it is important to clarify that they really are quite distinct from and do not help make the case for the community standard. Nevertheless, they are suggestive and worthy of examination, and we might well consider whether they have the ability to make equipoise extend longer, thus having relevance for the RCT dilemma.

One reason for interest in these other notions is the promise that they can be construed as being more relevant to clinical reality. The suggestion appears to be that we would avoid irrelevant hurdles by thinking about trials in a way that is more relevant to clinical care. They would then be seen not merely as a scheme to get equipoise to last longer so that we can carry out more trials, but as arising from an independent rationale.

Ultimately, though, the suggestion that these features would provide a way to extend equipoise further turns out to be misleading.

Now, there are various statements made by Freedman which can be construed in terms of seeing equipoise as clinical rather than theoretical, but I will spend the bulk of this discussion on what I take to be the single most plausible and significant one. As stated earlier, this is the interpretation that makes clinical vs. theoretical equipoise a distinction between *kinds of questions*: Clinical equipoise is equipoise about a practical question focused on clinical practice and the goal of healing, while *theoretical equipoise* is equipoise about a theoretical question focused on general scientific knowledge and the goal of understanding. Perhaps we would fall out of equipoise with respect to a theoretical question like 'Does treatment A cause outcome O with greater frequency (and extent) than does B?', while still remaining in equipoise with respect to a clinical question like 'Is A a better treatment for treating these patients than B is?'

As shown earlier in the paper, this (clinical question equipoise) is quite distinct from community equipoise, and thus will not help make the case for it. Still, it is worth evaluating in its own right.

15.8.1. Clinical Hypotheses as Multi-Demnsional Hypotheses

Freedman argues that theoretical equipoise concerns 'one-dimensional hypotheses', whereas real questions of clinical interest are more complex, involving multi-dimensional hypotheses (1987, p. 143). In the latter case, one wants to know whether a treatment is better than the alternative overall, that is, taking into account all its effects, positive and negative – the 'net therapeutic advantage' (1990, p. 5). This is, after all, what is clinically relevant. I think that this is the most significant and promising way of fleshing out this important idea of clinical as opposed to theoretical questions, both because this fits well with the idea of clinical relevance and because it points to a very suggestive argument (not given explicitly by Freedman) that can be offered for why clinical equipoise would keep us in equipoise longer.

So, how might this point about multi-dimensionality relate to reconceptualizing equipoise in a way that eases the RCT dilemma? Well, if we were only concerned with one dimension, a physician might well fall out of equipoise immediately or at least early on as the data begin to come in. But it will be pointed out that what's morally relevant is whether there is any evidence showing that the patient would be better off overall (with A rather than B), and this requires considering more than one dimension in the assessment. Thus perhaps one can reasonably stay in equipoise longer, perhaps throughout the trial (i.e., perhaps all the way to community equipoise, or even to an objective and reasonable measure of statistical significance).

Consider this in more detail: the choice between drugs A and B may be viewed as out of equipoise because, given a certain set of evidence, A is preferable with respect to outcome variable X (some measure of expected therapeutic benefit). However, at the same time, it might be that with respect to outcome variable Y (some side-effect), B is preferable to A. Suppose that B's superiority to A with respect to Y is just the right amount to offset A's superiority with respect to X. Then, considering both aspects (endpoints) at once, A and B *are* in (overall, or what we might call 'all dimensions considered' (ADC)) equipoise.

These facts might be used to argue as follows: 'Even though I am out of (individual) equipoise with respect to the main effect variable M, given that some side effect variable S goes the other way, I am actually in ADC (all dimensions considered) equipoise. And the latter is what really counts from the moral point of view.' In such a situation, the old (theoretical) view would have had us falling out of equipoise, whereas we now see that the right way to view it leaves us in equipoise, so the trial may surely be continued.

But this is a mistake. On the one hand, even if this sort of thing could move us some distance – allowing trials to go on under a legitimate sense of equipoise (clinical equipoise) even though theoretical equipoise had been lost – it is hard to see why it should be thought to move us sufficiently to get us to the point of community equipoise. In fact, however, a closer look suggests that it does not extend equipoise at all.

For if, with the main therapeutic endpoint M looking increasingly good, you are now in ADC equipoise, then before, with M being less positive, you would have been out of ADC equipoise in the other direction. So if you really were following the ADC interpretation of clinical equipoise, which, per hypothesis, is what really counts, you shouldn't have been able to start the trial.

Now, perhaps the situation envisaged is one where, as the interim data have come in, side effect S has gotten worse for arm A just as M has gotten better. Hence one really could have been in ADC equipoise throughout.

But, while this is true, it still will not help us. For if this is what is going on, then we do not have a case of mounting evidence showing the treatment to be ADC better, and thus we cannot learn over the course of this trial that arm A is better overall (ADC) than the comparison arm. We do not move any distance towards a point where we could be said to have sufficient evidence to make a policy decision about this kind of therapeutic action. So, just as, on the old view, we saw that we cannot reach statistical significance (about the theoretical hypothesis) without falling out of (theoretical, one-dimensional) equipoise along the way, we now see that we cannot move from a position of being in ADC equipoise to a point where the ADC difference between the two arms is large enough to be statistically significant without falling out of ADC equipoise along the way.

Another way to put all this is that, while it may well be appropriate and indeed preferable to change the endpoint followed to one which is more relevant to real clinical decisions, if we are to be consistent, we have to change that endpoint with respect to both what counts as equipoise (or what counts as personal care) and what counts as statistical significance (or enough evidence to attain the level of security of knowledge that was the point of the trial). That is, we must change it with respect to both (a) what counts as 'crossing the line' with respect to the *individual-patient decision*, and (b) what counts as 'crossing the line' with respect to the *policy decision*. And once we change both criteria to that endpoint, there is still (or again) a wide gap between equipoise and statistical significance (or individual-case-acceptable evidence and policy-acceptable evidence). As a result, clinical equipoise in the sense of multi-dimensionality is not going to help us to resolve the RCT dilemma.[8]

15.8.2. Inherent Imprecision

As described above, Freedman's discussion of the clinical side of clinical equipoise includes some features besides multi-dimensionality. In particular, much of the argument using the language of 'clinical relevance' focuses instead on how individual judgments are often idiosyncratic and sensitive to such 'artifacts' as the particular modes of measurement and representation relied upon (e.g., which way data are plotted) (1987, p. 143). Presumably there is no one such mode of representation which is perfect and free of assumptions, and so there is necessarily a certain amount of 'noise' or inherent imprecision.

But while it is appropriate and significant to pay attention to these sources of noise or inherent imprecision, and it also seems fair enough to say that this is 'part of clinical reality', it is not accurate to assert that these features are not present when scientific or theoretical questions are pursued, and so there will be a similar decrease in our confidence in these contexts as well. So whatever lengthening of equipoise arises form such noise or imprecision is not a function of moving from the theoretical to a clinical question.

And even if this were not the case, the suggestion that this process will lengthen the time during which we are in equipoise in a way that will aid in resolving the RCT dilemma is mistaken. Granted, this sort of decrease in confidence might extend the time during which we are truly undecided about what is best for our present patient. But it will equally extend the time during which we are truly undecided about the *policy* decision (e.g., the decision that it is time to stop the trial and put the drug on the market). So there will be the same gap between evidence sufficient for a present patient decision and evidence sufficient for a policy decision.

To sum up these matters concerning clinical equipoise: the notion of clinical (as opposed to theoretical) equipoise is itself ambiguous. But neither in the sense of imprecision nor in the (more profound) sense of multi-dimensionality does it provide a criterion that helps to resolve the RCT dilemma.

Concerning the multi-dimensionality aspect of clinical equipoise: There may well be reason to argue that trials should be designed around clinical rather than theoretical questions. But having done so – and having chosen a question – one still has to go from equipoise through partial knowledge to statistical significance. The move to a clinical question doesn't change this basic fact.

Concerning the matters of artifacts and imprecision: These are indeed processes that increase the difficulty of making a precise assessment. But, again, this doesn't really get around the above facts about the gradualness of our attainment of security of knowledge and how present-patient and policy decision-contexts require different degrees of such evidence.

In addition, both the clinical question idea and the sensitivity to artifacts idea are distinct from community equipoise, and neither one provides an argument that would provide support for that criterion. Finally, as argued in the earlier part of the paper, there are good reasons to deny both the legitimacy of following community equipoise and its sufficiency in allowing us to carry out a trial to the relevant point.

I conclude that the various notions proposed in the work of Freedman and others concerning what they label 'clinical equipoise' do not help us resolve the RCT dilemma.

15.9. FINAL COMMENTS

Much of the problem here has arisen from the failure to keep in mind the fact that there are a number of different senses of equipoise. In fact, the concept of equipoise is even more ambiguous than described here. A full analysis would require making a number of other distinctions. For instance, one important such distinction concerns the question of who needs to be in equipoise: patients or clinicians. Some others are mentioned in (Gifford 2000, p. 423). The task here has been the limited one of subjecting this particular influential proposal of Freedman's to critical scrutiny, so we have left these other questions aside.

Finally, the goal here is not to argue that we should stop doing RCTs. It is to say that we should not rest comfortably with the idea that we have, in the concept of clinical equipoise, a justification for and an adequate understanding or concep-tualization of the moral issues involved. Freedman's account might have appeared to provide a special line to be drawn, about what counts as 'really knowing,' that can be used to resolve the RCT dilemma. But it does not do so. The failure of this account would appear to require us to rely substantially more on either or both of the following. We can work much harder to ensure the truly voluntary, truly informed consent of all participants in clinical trials. We can face more squarely the value trade-offs intrinsic to clinical trials, opening up such questions as: How much knowledge is worth what in terms of suboptimal treatment for present patient-subjects? And how sure do we have to be that subjects giving informed consent understand the complexities of trials? These questions are themselves as important as they are both complicated and disturbing, but they are beyond the scope of the present paper.

Michigan State University, East Lansing, Michigan, USA

NOTES

[1] This way of describing the matter relies on the notion of statistical significance as understood in standard frequentist statistics. Bayesians will think that the evidence increases continuing through the trial and that there may be compelling evidence that a treatment works long before statistical significance is reached. There may be good reasons for adopting a Bayesian analysis of clinical trial design and interpretation. However, the moral dilemma discussed here, caused by the gradually evolving amounts of evidence, remains no matter which way this is described.

[2] Cf. (Rudner, 1953) for an argument that this applies to the acceptance of scientific hypotheses in general, and thus requires value judgments in science.

[3] It might be worth considering some parallels here with debates about and resistance to clinical practice guidelines, where there are also complaints about merely applying the judgments of others (Boyle, 1998).

[4] Each of these dimensions could perhaps be filled in a more fine-grained way; perhaps each could be seen as a continuum. Miller and Weijer (2003) have recently proposed something short of unanimity, what might be called the 'respectable minority' view, namely that equipoise remains just in case a 'respectable minority' remains unconvinced of the major view.

[5] Note that there is a parallel here with the topics of placebo-controlled AZT trials in developing nations, where there has been debate over whether it is relevant or sufficient to argue that the subjects given a placebo would not have had access to the study drug in any case (Varmus & Satcher, 1997; Lurie & Wolfe, 1997).

[6] That this is so is more explicit in (Freedman, 1990) and (Freedman, 1992).

[7] Put otherwise, if we can assume people will change their minds simultaneously in this way, we have a direct solution to the RCT dilemma and don't have to resort to the views of the community.

[8] In (Gifford, 2000), I develop a line of thought utilizing some of these facts about multi-dimensionality to attempt to resolve the dilemma by taking advantage of such matters as the differences in values between different subjects, noting that different subjects will be in equipoise at different times. But I also show that insofar as this works as a solution, it is quite distinct from the clinical or community equipoise models proposed by Freedman.

REFERENCES

Boyle P (1998) Getting doctors to listen: ethics and outcomes data in context. Georgetown University Press, Washington, D.C.

Crouch R, Arras J (1998) AZT Trials and Tribulations. Hastings Center Report 28(6):26–34

Freedman B (1987) Equipoise and the ethics of clinical research. N Engl J Med 317:141–145

Freedman B (1990) Placebo-controlled trials and the logic of clinical purpose. IRB 12(6):1–6

Freedman B (1992) A response to a purported ethical difficulty with randomized clinical trials involving cancer patients. J Clin Ethics 3:231–234

Fried C (1974) Medical experimentation: Personal integrity and social policy. North-Holland Publishing, Amsterdam

Gifford F (1986) The conflict between randomized clinical trials and the therapeutic obligation. J Med Philos 11:347–366

Gifford F (1995) Community equipoise and the ethics of randomized clinical trials. Bioethics 9:127–148

Gifford F (2000) Freedman's clinical equipoise and sliding-scale all-dimensions considered equipoise. J Med Philos 25:399–426

Hellman S, Hellman DS (1991) Of mice but not men. N Engl J Med 324:1585–1589

Lurie P, Wolfe S (1997) Unethical trials of interventions to reduce perinatal transmission of the human immunodeficiency virus in developing countries. N Engl J Med 337:853–856

Marquis D (1983) Leaving therapy to chance. Hastings Center Report 13(4):40–47

Miller P, Weijer C (2003) Rehabilitating Equipoise. Kennedy Institute of Ethics Journal 13:93–118

Rudner R (1953) The scientist qua scientist makes value judgments. Philosophy of Science 20:1–6

Weijer, Shapiro, Glass, Enkin (2000) For and against: Clinical equipoise and not the uncertainty principle is the moral underpinning of the randomized controlled trial. Br Med J 321:756–758

Varmus H, Satcher D (1997) Ethical complexities of conducting research in developing countries. N Engl J Med 337:1003–1005

INDEX

235

Philosophy and Medicine

1. H. Tristram Engelhardt, Jr. and S.F. Spicker (eds.): *Evaluation and Explanation in the Biomedical Sciences.* 1975 ISBN 90-277-0553-4
2. S.F. Spicker and H. Tristram Engelhardt, Jr. (eds.): *Philosophical Dimensions of the Neuro-Medical Sciences.* 1976 ISBN 90-277-0672-7
3. S.F. Spicker and H. Tristram Engelhardt, Jr. (eds.): *Philosophical Medical Ethics.* Its Nature and Significance. 1977 ISBN 90-277-0772-3
4. H. Tristram Engelhardt, Jr. and S.F. Spicker (eds.): *Mental Health.* Philosophical Perspectives. 1978 ISBN 90-277-0828-2
5. B.A. Brody and H. Tristram Engelhardt, Jr. (eds.): *Mental Illness.* Law and Public Policy. 1980 ISBN 90-277-1057-0
6. H. Tristram Engelhardt, Jr., S.F. Spicker and B. Towers (eds.): *Clinical Judgment.* A Critical Appraisal. 1979 ISBN 90-277-0952-1
7. S.F. Spicker (ed.): *Organism, Medicine, and Metaphysics.* Essays in Honor of Hans Jonas on His 75th Birthday. 1978 ISBN 90-277-0823-1
8. E.E. Shelp (ed.): *Justice and Health Care.* 1981 ISBN 90-277-1207-7; Pb 90-277-1251-4
9. S.F. Spicker, J.M. Healey, Jr. and H. Tristram Engelhardt, Jr. (eds.): *The Law-Medicine Relation.* A Philosophical Exploration. 1981 ISBN 90-277-1217-4
10. W.B. Bondeson, H. Tristram Engelhardt, Jr., S.F. Spicker and J.M. White, Jr. (eds.): *New Knowledge in the Biomedical Sciences.* Some Moral Implications of Its Acquisition, Possession, and Use. 1982 ISBN 90-277-1319-7
11. E.E. Shelp (ed.): *Beneficence and Health Care.* 1982 ISBN 90-277-1377-4
12. G.J. Agich (ed.): *Responsibility in Health Care.* 1982 ISBN 90-277-1417-7
13. W.B. Bondeson, H. Tristram Engelhardt, Jr., S.F. Spicker and D.H. Winship: *Abortion and the Status of the Fetus.* 2nd printing, 1984 ISBN 90-277-1493-2
14. E.E. Shelp (ed.): *The Clinical Encounter.* The Moral Fabric of the Patient-Physician Relationship. 1983 ISBN 90-277-1593-9
15. L. Kopelman and J.C. Moskop (eds.): *Ethics and Mental Retardation.* 1984 ISBN 90-277-1630-7
16. L. Nordenfelt and B.I.B. Lindahl (eds.): *Health, Disease, and Causal Explana-tions in Medicine.* 1984 ISBN 90-277-1660-9
17. E.E. Shelp (ed.): *Virtue and Medicine.* Explorations in the Character of Medicine. 1985 ISBN 90-277-1808-3
18. P. Carrick: *Medical Ethics in Antiquity.* Philosophical Perspectives on Abortion and Euthanasia. 1985 ISBN 90-277-1825-3; Pb 90-277-1915-2
19. J.C. Moskop and L. Kopelman (eds.): *Ethics and Critical Care Medicine.* 1985 ISBN 90-277-1820-2
20. E.E. Shelp (ed.): *Theology and Bioethics.* Exploring the Foundations and Frontiers. 1985 ISBN 90-277-1857-1
21. G.J. Agich and C.E. Begley (eds.): *The Price of Health.* 1986 ISBN 90-277-2285-4
22. E.E. Shelp (ed.): *Sexuality and Medicine.* Vol. I: Conceptual Roots. 1987 ISBN 90-277-2290-0; Pb 90-277-2386-9
23. E.E. Shelp (ed.): *Sexuality and Medicine.* Vol. II: Ethical Viewpoints in Transition. 1987 ISBN 1-55608-013-1; Pb 1-55608-016-6
24. R.C. McMillan, H. Tristram Engelhardt, Jr., and S.F. Spicker (eds.): *Euthanasia and the Newborn.* Conflicts Regarding Saving Lives. 1987 ISBN 90-277-2299-4; Pb 1-55608-039-5

Philosophy and Medicine

25. S.F. Spicker, S.R. Ingman and I.R. Lawson (eds.): *Ethical Dimensions of Geriatric Care.* Value Conflicts for the 21th Century. 1987   ISBN 1-55608-027-1
26. L. Nordenfelt: *On the Nature of Health.* An Action-Theoretic Approach. 2nd, rev. ed. 1995   ISBN 0-7923-3369-1; Pb 0-7923-3470-1
27. S.F. Spicker, W.B. Bondeson and H. Tristram Engelhardt, Jr. (eds.): *The Contraceptive Ethos.* Reproductive Rights and Responsibilities. 1987   ISBN 1-55608-035-2
28. S.F. Spicker, I. Alon, A. de Vries and H. Tristram Engelhardt, Jr. (eds.): *The Use of Human Beings in Research.* With Special Reference to Clinical Trials. 1988
                                    ISBN 1-55608-043-3
29. N.M.P. King, L.R. Churchill and A.W. Cross (eds.): *The Physician as Captain of the Ship.* A Critical Reappraisal. 1988   ISBN 1-55608-044-1
30. H.-M. Sass and R.U. Massey (eds.): *Health Care Systems.* Moral Conflicts in European and American Public Policy. 1988   ISBN 1-55608-045-X
31. R.M. Zaner (ed.): *Death: Beyond Whole-Brain Criteria.* 1988   ISBN 1-55608-053-0
32. B.A. Brody (ed.): *Moral Theory and Moral Judgments in Medical Ethics.* 1988
                                    ISBN 1-55608-060-3
33. L.M. Kopelman and J.C. Moskop (eds.): *Children and Health Care.* Moral and Social Issues. 1989   ISBN 1-55608-078-6
34. E.D. Pellegrino, J.P. Langan and J. Collins Harvey (eds.): *Catholic Perspectives on Medical Morals.* Foundational Issues. 1989   ISBN 1-55608-083-2
35. B.A. Brody (ed.): *Suicide and Euthanasia.* Historical and Contemporary Themes. 1989
                                    ISBN 0-7923-0106-4
36. H.A.M.J. ten Have, G.K. Kimsma and S.F. Spicker (eds.): *The Growth of Medical Knowledge.* 1990   ISBN 0-7923-0736-4
37. I. Löwy (ed.): *The Polish School of Philosophy of Medicine.* From Tytus Chałubiński (1820–1889) to Ludwik Fleck (1896–1961). 1990   ISBN 0-7923-0958-8
38. T.J. Bole III and W.B. Bondeson: *Rights to Health Care.* 1991   ISBN 0-7923-1137-X
39. M.A.G. Cutter and E.E. Shelp (eds.): *Competency.* A Study of Informal Compe-tency Determinations in Primary Care. 1991   ISBN 0-7923-1304-6
40. J.L. Peset and D. Gracia (eds.): *The Ethics of Diagnosis.* 1992   ISBN 0-7923-1544-8
41. K.W. Wildes, S.J., F. Abel, S.J. and J.C. Harvey (eds.): *Birth, Suffering, and Death.* Catholic Perspectives at the Edges of Life. 1992 [CSiB-1]
                                    ISBN 0-7923-1547-2; Pb 0-7923-2545-1
42. S.K. Toombs: *The Meaning of Illness.* A Phenomenological Account of the Different Perspectives of Physician and Patient. 1992   ISBN 0-7923-1570-7; Pb 0-7923-2443-9
43. D. Leder (ed.): *The Body in Medical Thought and Practice.* 1992 ISBN 0-7923-1657-6
44. C. Delkeskamp-Hayes and M.A.G. Cutter (eds.): *Science, Technology, and the Art of Medicine.* European-American Dialogues. 1993   ISBN 0-7923-1869-2
45. R. Baker, D. Porter and R. Porter (eds.): *The Codification of Medical Morality.* Historical and Philosophical Studies of the Formalization of Western Medical Morality in the 18th and 19th Centuries, Volume One: Medical Ethics and Etiquette in the 18th Century. 1993   ISBN 0-7923-1921-4
46. K. Bayertz (ed.): *The Concept of Moral Consensus.* The Case of Technological Inter-ventions in Human Reproduction. 1994   ISBN 0-7923-2615-6
47. L. Nordenfelt (ed.): *Concepts and Measurement of Quality of Life in Health Care.* 1994 [ESiP-1]   ISBN 0-7923-2824-8

Philosophy and Medicine

Philosophy and Medicine

70. A. Nordgren: *Responsible Genetics*. The Moral Responsibility of Geneticists for the Consequences of Human Genetics Research. 2001 ISBN 1-4020-0201-7
71. J. Tao Lai Po-wah (ed.): *Cross-Cultural Perspectives on the (Im)Possibility of Global Bioethics*. 2002 [ASiB-2] ISBN 1-4020-0498-2
72. P. Taboada, K. Fedoryka Cuddeback and P. Donohue-White (eds.): *Person, Society and Value*. Towards a Personalist Concept of Health. 2002 ISBN 1-4020-0503-2
73. J. Li: *Can Death Be a Harm to the Person Who Dies?* 2002 ISBN 1-4020-0505-9
74. H.T. Engelhardt, Jr. and L.M. Rasmussen (eds.): *Bioethics and Moral Content: National Traditions of Health Care Morality*. Papers dedicated in tribute to Kazu-masa Hoshino. 2002 ISBN 1-4020-6828-2
75. L.S. Parker and R.A. Ankeny (eds.): *Mutating Concepts, Evolving Disciplines: Genetics, Medicine, and Society*. 2002 ISBN 1-4020-1040-0
76. W.B. Bondeson and J.W. Jones (eds.): *The Ethics of Managed Care: Professional Integrity and Patient Rights*. 2002 ISBN 1-4020-1045-1
77. K.L. Vaux, S. Vaux and M. Sternberg (eds.): *Covenants of Life. Contemporary Medical Ethics in Light of the Thought of Paul Ramsey*. 2002 ISBN 1-4020-1053-2
78. G. Khushf (ed.): *Handbook of Bioethics: Taking Stock of the Field from a Philo-sophical Perspective*. 2003 ISBN 1-4020-1870-3; Pb 1-4020-1893-2
79. A. Smith Iltis (ed.): *Institutional Integrity in Health Care*. 2003 ISBN 1-4020-1782-0
80. R.Z. Qiu (ed.): *Bioethics: Asian Perspectives A Quest for Moral Diversity*. 2003 [ASiB-3] ISBN 1-4020-1795-2
81. M.A.G. Cutter: *Reframing Disease Contextually*. 2003 ISBN 1-4020-1796-0
82. J. Seifert: *The Philosophical Diseases of Medicine and Their Cure*. Philosophy and Ethics of Medicine, Vol. 1: Foundations. 2004 ISBN 1-4020-2870-9
83. W.E. Stempsey (ed.): *Elisha Bartlett's Philosophy of Medicine*. 2004 [CoME-2] ISBN 1-4020-3041-X
84. C. Tollefsen (ed.): *John Paul II's Contribution to Catholic Bioethics*. 2005 [CSiB-3] ISBN 1-4020-3129-7
85. C. Kaczor: *The Edge of Life*. Human Dignity and Contemporary Bioethics. 2005 [CSiB-4] ISBN 1-4020-3155-6
86. R. Cooper: *Classifying Madness*. A Philosophical Examination of the Diagnostic and Statistical Manual of Mental Disorders. 2005 ISBN 1-4020-3344-3
87. L. Rasmussen (ed.): *Ethics Expertise*. History, Contemporary Perspectives, and Applications. 2005 ISBN 1-4020-3819-4
88. M.C. Rawlinson and S. Lundeen (eds.): *The Voice of Breast Cancer in Medicine and Bioethics*. 2006 ISBN 1-4020-4508-5
89. M. Bormuth: *Life Conduct in Modern Times*. Karl Jaspers and Psychoanalysis. 2006 ISBN 1-4020-4764-9
90. H. Kincaid and J. McKitrick (eds.): *Establishing Medical Reality*. Essays in the Metaphysics and Epistemology of Biomedical Science. 2007 ISBN 1-4020-5215-4
91. S.C. Lee (ed.): *The Family, Medical Decision – Making, and Biotechnology*. Critical Reflections on Asian Moral Perspectives. 2007 ISBN 1-4020-5219-7
92. H.G. Wright: *Means, Ends and Medical Care*. 2007 ISBN 1-4020-5219-X